ANGELS, DEVIL AND SCIENCE

A collection of articles on scientific temper

Popular Science

ANGELS, DEVIL AND SCIENCE

A collection of articles on scientific temper

Pushpa M Bhargava
and
Chandana Chakrabarti

nbt.india
एकः सूते सकलम्
NATIONAL BOOK TRUST, INDIA

Cover photograph courtesy: www.worth1000.com

ISBN 978-81-237-5182-5

First Edition 2007
Reprints 2009, 2010, 2012, 2016, 2018, 2021, 2023, 2025 (*Saka* 1947)

₹ 245.00

Published by the Director, National Book Trust, India
5 Institutional Area, Vasant Kunj, New Delhi- 110 070
Website : www.nbtindia.gov.in

To

RITI and DIPON

For their total commitment to Scientific Temper

Contents

Acknowledgements

We are grateful to K Mohan and Chandra Prakash for typing the manuscript, and to Ganga Nair, G R Ashok Kumar and Vinod Kumar for office support.

Abbreviations

(Many of the abbreviations used have also been defined in the text)

AP	Andhra Pradesh
ASWI	Association of Scientific Workers of India
CC	Chandana Chakrabarti
CCMB	Centre for Cellular and Molecular Biology, Hyderabad
CFTRI	Central Food Technological Research Institute, Mysore
CSIR	Council of Scientific and Industrial Research
DNA	Deoxyribonucleic acid (the genetic material)
IICT	Indian Institute of Chemical Technology, Hyderabad
INSA	Indian National Science Academy
IRA	Indian Rationalists Association
ISRO	Indian Space Research Organisation
MOSE	The Method of Science Exhibition
NCERT	National Council for Educational Research and Training
NCSTC	National Council for Science and Technology Communication
PM	Prime Minister
PMB	Pushpa M Bhargava
RNA	Ribonucleic acid
RRLH	Regional Research Laboratory, Hyderabad
RSS	Rashtriya Swayamsevak Sangh

UAE	United Arab Emirates
UK	United Kingdom
USA/US	United States of America
WMD	Weapons of Mass Destruction

('he' and 'him', wherever used, include 'she' and 'her', respectively.)

Note: Bombay is now called Mumbai
Calcutta is now called Kolkata
Ceylon is now called Sri Lanka

I

WHAT IS IT ALL ABOUT?

A STATEMENT BY 192 LEADING SCIENTISTS*

Scientists in a variety of fields have become concerned about the increased acceptance of astrology in many parts of the world. We, the undersigned—astronomers, astrophysicists, and scientists in other fields—wish to caution the public against the unquestioning acceptance of the predictions and advice given privately and publicly by astrologers. Those who wish to believe in astrology should realize that there is no scientific foundation for its tenets.

In ancient times people believed in the predictions and advice of astrologers because astrology was part and parcel of their magical world view. They looked upon celestial objects as abodes or omens of the Gods and, thus, intimately connected with events here on earth; they had no concept of the vast distances from the earth to the planets and stars. Now that these distances can and have been calculated, we can see how infinitesimally small are the gravitational and other effects produced by the distant planets and the far more distant stars. It is simply a mistake to imagine that the forces exerted by stars and planets at the moment of birth can in any way shape our futures. Neither is it true that the position of distant heavenly bodies make certain days or periods more favourable to particular kinds of action, or that the sign under which one

* *The Humanist*, Sept/Oct 1975.
19 Nobel Prize Winners and 173 distinguished scientists from a variety of fields from all over the world had signed the above statement in 1975.

was born determines one's compatibility or incompatibility with other people.

Why do people believe in astrology? In these uncertain times many long for the comfort of having guidance in making decisions. They would like to believe in a destiny predetermined by astral forces beyond their control. However, we must all face the world, and we must realize that our futures lie in ourselves, and not in the stars.

One would imagine, in this day of widespread enlightenment and education, that it would be unnecessary to debunk beliefs based on magic and superstition. Yet, acceptance of astrology pervades modern society. We are especially disturbed by the continued uncritical dissemination of astrological charts, forecasts, and horoscopes by the media and by otherwise reputable newspapers, magazines, and book publishers. This can only contribute to the growth of irrationalism and obscurantism. We believe that the time has come to challenge directly, and forcefully, any pretentious claims of astrological charlatans.

It should be apparent that those individuals who continue to have faith in astrology do so in spite of the fact that there is no verified scientific basis for their beliefs, and indeed that there is strong evidence to the contrary.

* * *

THE END OF HOMOEOPATHY*

That homeopathy fares poorly when compared with allopathy in Aijing Shang and colleagues' systematic evaluation is unsurprising. Of greater interest is the fact that this debate continues, despite 150 years of unfavourable findings. The more dilute the evidence for homoeopathy becomes, the greater seems its popularity.

For long, a politically correct laissez-faire attitude has existed towards homoeopathy, but there are now signs of enlightenment from unlikely sources. The UK Parliamentary Select Committee on Science & Technology issued a report about complementary and alternative medicine in 2000. It recommended "any therapy that makes specific claims for being able to treat specific conditions should have evidence of being able to do this above and beyond the placebo effect". Going one step further, the Swiss Government, after a 5-year trial, has now withdrawn insurance coverage for homoeopathy and four other complementary treatments because they did not meet efficacy and cost-effectiveness criteria.

In a Comment (*Lancet*, this issue), Jan Vandenbroucke gives a philosophical interpretation of Shang's study. Other philosopher he might have included is Kant, who reminds us that we see things not as they are, but as we are. This observation is also true of health-care consumers, who may see homoeopathy as a holistic alternative to a disease-focused, technology-driven medical model. It is the attitudes of patients and providers that engender alternative-therapy seeking behaviours which create a greater threat to conventional care—and patients' welfare—than do spurious arguments of putative benefits from absurd dilutions.

Surely the time has passed for selective analyses, biased reports, or further investment in research to perpetuate the homoeopathy versus allopathy debate. Now doctors need to be bold and honest with their patients about homeopathy's lack of benefit, and with themselves about the failings of modern medicine to address patients' needs for personalized care.

* * *

* *The Lancet*, 27th August 2005, Vol 366, p.690, Editorial

WOMAN RAPED AT PUTTAPARTHI*

Besler Karia, a German national has lodged a police complaint alleging she was repeatedly raped at Sai Nilayam apartments in Puttaparthi. In her complaint lodged on August 29, the victim also alleged that she was drugged and subjected to repeated rape since her arrival in the first week of July. A case was registered to this effect by Puttaparthi police, crime number 100/2005. She was referred to the Anantapur general hospital where she would be kept under observation for two days.

Puttaparthi DSP, Mr Krishnamurthy, seems sceptical about her allegations. He said Karia was mentally disturbed and was addicted to drugs. The DSP claimed the German national was not making a coherent statement regarding the crime. However, sources in Anantapur general hospital told this correspondent that the victim wanted her blood samples to be taken to verify whether she was administered drugs or not.

The Puttaparthi CI has referred the case to the Anantapur general hospital for vaginal smear tests and other examinations required to confirm the crime. Sources stated that the victim named some individuals, but the names were kept a secret. However the victim refused to talk to media persons.

A similar case was witnessed in Puttaparthi a few months back when an Ethiopian passport holder, Zenneth had alleged she had been subjected to repeated rape during her visit to the city. However, her contention too was brushed aside as rantings of a mentally imbalanced individual. No formal complaint had been lodged in this lady's case. In several letters written to this newspaper she had complained of being raped by locals.

The case of Ms Besler Karia is the first of its kind wherein a formal complaint of rape was lodged with the police. Puttaparthi, is considered a sacred place; most negative incidents are quickly hushed up lest it should have an adverse impact on the sanctity of the place. Puttaparthi DSP, Mr Krishnamurthy, has dismissed the contention of the German

* *Deccan Chronicle*, Hyderabad, 1st September 2005

national as ridiculous. She says she was continuously raped ever since she came here. In her complaint the German national has also stated that she was raped at Chennai where she had gone to discuss this matter with the German Consulate. "The allegation is hard to believe. Nevertheless we have referred the case to Anantapur general hospital", he says. He, however, made it clear that he was unaware of the case related to Ms Zenneth.

Just asking

- Why is it that every unpleasant incident in Puttaparthi is hushed up and cops often feign ignorance?
- Time and again one has heard of 'unholy' activities in Puttaparthi, but it is alleged the Andhra Pradesh government never found it necessary to find out why no FIR (First Information Report) has been filed.
- Two women have filed complaints, saying they've been raped, and the cops dismiss them both as "mentally unsound". Does this mean only unsound people go to Puttaparthi?

* * *

POLICE TO BATTLE 'BANAMATI' DEATHS*

This year, 54 people were killed on charge of practising witchcraft.

The sudden spurt in killing of people in the name of practising *banamati* (witchcraft) in Telangana districts has spurred the police to join hands with non-governmental organizations (NGOs) to campaign against the inhuman act.

The year saw 54 persons being killed on charge of practising witchcraft. In the last fortnight, five persons were battered, lynched and set ablaze for allegedly practising black magic. Be it Mudireddipalli (Mahabubnagar), Mandapuram (Nalgonda), Narsapur (Medak)or Nizamabad district, suspected persons were tortured and brutally killed.

Those practising sorcery were mercilessly killed for ills plaguing the families or the entire village.

Shocked by the daylight murder of a Dalit woman on the charge of performing black magic in Nampally mandal on Tuesday, the Nalgonda police decided to launch a joint campaign against *banamati* with an NGO, Jana Vignana Vedika (JVV). Asserting that the police would not allow such incidents, SP M M Bhagwat warned perpetrators that they would not be spared.

JVV general secretary T V Rao said massive awareness campaign would be launched in all the nine Telangana districts with the support of police, Government machinery and political parties.

The JVV has decided to write to all SPs in Telangana and political parties to extend support to their cause wherein scientists and magicians would be roped in.

Mr Rao said Nalgonda was witnessing rise in such cases and pointed out that mob-fury had claimed 54 lives this year alone in the State due to low literacy levels and economic backwardness.

Land disputes, family squabbles, rivalry between two

* *The Hindu*, Hyderabad, 14th October 2005

groups were also shown as reasons for persons getting killed in the name of witchcraft, he noted.

In many of these cases, irate mobs plucked the teeth of victims in the belief that it would disable them from chanting hymns to invoke evil spirits.

Social boycott

Burning of houses and social boycott of so-called sorcerers are not uncommon.

In the absence of qualified doctors, quacks are resorting to sorcery and witchcraft to cure ailments of villagers, quite often with fatal consequences for both.

* * *

TOWN DWELLERS APPEASE 'EVIL FORCES'*

Town residents are as prone to superstition as are people of interior villages.

Residents of Shantinagar Colony (Ward No.27) of Nirmal town and sorcerers from Karimnagar performed *kshudrapujalu* (occult practices) and sacrificed goats and hens on Sunday to appease "evil forces" and bring peace, following the increased number of deaths of youth and married men over the last three years in the colony.

Although the deaths were more in families of the Vaddara community, people from all communities participated in the event in the interests of the colony. They spent nearly Rs.50,000 for the event.

Residents of the colony and sorcerers were busy preparing for the event over the last three days. During the *kshudrapujalu*, nobody from the colony was allowed to go outside and nobody was allowed into the colony on the directions of the sorcerer. Four goats were sacrificed and their blood was splashed on the four routes of the colony. Hens were sacrificed in the center of the colony to protect residents from evil forces and bring peace.

In another instance, villagers of Manjulapur of Nirmal mandal performed *kshudrapujalu* and sacrificed goats and hens in the late hours of Monday to appease evil forces, following the deaths of cattle in their villages.

Such incidents have also occurred in Kuntla, Dilawarpur, Laxmanchanda and Mamada mandal in Nirmal division in the past, where villagers deserted villages.

Residents said that nearly 20 persons, most of them youth and newly married persons, died in suspicious circumstances and fluke incidents over the last three years. A sorcerer, Devatha Rajanna, 50, of Vempeta of Karimnagar said that they had agreed to perform *kshudrapujalu* to save the residents from evil forces. He said that there was nothing wrong in performing such *pujas* when people strongly believed superstitions even in these modern times.

* * *

* *Deccan Chronicle*, Hyderabad, 25th October 2005

A LOST VEDIC RITUAL IS BROUGHT ALIVE*

The ancient ritual of Ashwamedha Yagnam, conducted by emperors to extend their domains, will be resurrected in Hyderabad on December 11.

In the modern age, however, the ritual is intended to extend the domain of peace and tranquility across the world.

The Hyderabad-based Aananda Aashramam will conduct the Ashwamedha Yagnam and will invite people of all religions to take part in it and offer prayers for the well-being of humanity.

It was a popular Vedic ritual in ancient India but became rare in the last 2,000 years.

"It is a once-in-a-lifetime opportunity," said Dr P V Sesha Sai, the *yaga kartha* or organizer.

Ashwamedha Yagnam is of two kinds—Snarta Ashwamedha Yagnam and Srouta Ashwamedha Yagnam. Kings used to perform Srouta Yagnam and would sacrifice a horse at the end of the ritual. The Snarta Yagnam was performed by saints without animal sacrifice.

"We are going to observe the Snarta variety in Hyderabad, reviving a tradition which became extinct long ago," said Dr Sesha Sai.

It is a Vedic ritual which has a deeply secular and political dimensions. The role of the priest is not as prominent as in other rituals.

The *sankalpa* for the yagnam took place in the month of Maha Maghi (February). "We have selected a bright day of Margasira *maas* coupled with Sunday and Revati Star to hold the yagnam," said Dr Sai. "It will be on for eight days."

Jagadguru Sri Ganeshananda Bharathi Mahaswami will bless the yagnam which will be performed as per Vedic traditions.

Along with Ashwamedha Yagnam, priests will also conduct other yagnams. Individuals aspiring for "political prosperity" can take part in Sri Sowra Yagnam. Sri Varuna Yagnam will

* *Deccan Chronicle*, Hyderabad, 26th October 2005

bring timely rains and Sri Runa Vimochana Pasupatam will help a person escape from debt traps.

The priests will also conduct Sri Kanya Pasupatam (for marriage), Sri Kubera Pasupatam (for business success). Sri Mrutyunjaya Homam (for long life), and Sri Navagraha Pasupatam (for relief from bad times), along with the main yagnam.

Dr Sesha Sai said that there was a clear difference between Ashwamedha (horse sacrifice) which is held as an 'outward' rite and the one held as an 'inward' one.

The external rite was performed by kings.

However, the horse is also the symbol for life-energy inside the human being and in the cosmos. *Medha* means 'offering' as well as 'intelligence'. In the 'inward rite', Ashwamedha is an offering of the life force to the deity.

As part of the ritual, 108 couples, 108 students, 108 housewives and 108 girls will be honoured. In tune with the times, a Sarva Dharma Sammelan will also be held.

The yagnam will be held in about 100 acres of land and 300 Vedic scholars will participate in it. And of course, horses will be very much a part of the ritual.

* * *

'VISA' TEMPLES BECOMING A RAGE IN AP*

In a state where seeking to migrate abroad is fast becoming an obsession, is it surprising to find "Visa temples"? A second visa temple has recently come up in Andhra Pradesh in an obscure village, 45 km from Nizamabad town in backward Telangana.

Hundreds of visa seekers are thronging a dead 100-year old banyan tree in Mupkal village, after the successful ones spread word about its power. Most of them visiting are skilled labourers like motor mechanics keen on getting a job in Arab countries. They come and tie a coconut wrapped in a red or white cloth and promise to keep a vow.

This is the second visa boom granting temple, the first one being the "Visa God" at Chilkur Balaji temple in Himayatsagar on the outskirts of Hyderabad. Thousands come to worship here and have been doing for some years. This dead banyan tree in Nizamabad which has now been converted into a makeshift temple is getting hundreds of devotees from various parts of the district, including Nizamabad town, Warni, Vannel, Kammarpalli, Kottaplli, Venkatapur and Asakottur since June 24 this year.

The local priest at the makeshift temple, N. Nagalingam, told *The Times of India* that the list of pilgrims has been on the rise ever since eight persons who worshipped at the place got a visa to go to West Asia last month. "It's a mela here every day. Several petty vendors have set up shop in the vicinity to cater to the hundreds of devotees who come here with a wish list, most of them on Mondays and Saturdays", says Nagalingam.

Locals said that the tree, more than 100 years old, began dying and almost got uprooted on January 28 due to a hailstorm. After the incident, villagers decided to cut down the tree as it posed a threat to residents in the vicinity. A major portion of the tree was cut with only a huge trunk extending to about 15 feet remaining in the place.

* *The Times of India*, Bangalore, 2005

On June 24, a heavy storm accompanied by gale restored the nearly uprooted trunk to its earlier position, thereby "convincing" villagers that the dead tree was blessed with unnatural powers. "Since that day, pilgrims have started coming," says the priest. A motor mechanic from Mupkal village, P Mahesh, said, "I got a visa to Dubai (UAE) recently after worshipping here. I am flying off in September."

Despite an awareness drive by Jana Vignana Vedika State representative, Ch Rajeshwar, that the heavy trunk came back to its original position following heavy rains and the law of gravity, the list of pilgrims seems increasing by the day.

* * *

BIG B KEEPS HIS PROMISE*

Living up to his promise, Bollywood star Amitabh Bachchan, sent a jewellery experts' team from Mumbai for taking measurements of 'Kati Abhaya Hastams' of the presiding deity at the Tirumala temple.

According to temple officials the team, whose identity has not been revealed, reached here on Sunday and secured the measurements the same evening.

It may be recalled that Big B, who came on a thanksgiving visit during the last week of December after surviving a major surgery, had promised the temple authorities to donate Rs.8 crores for manufacturing twin diamond-studded gold *hastams*.

But, according to highly placed sources, the *hastams* are expected to cost Rs.12 crores though initial estimates had put the cost at Rs.8 crores.

* * *

* *The Hindu*, Hyderabad, 17th January 2006

POLITICIANS IN TROUBLE SEEK TANTRIK HELP*

Seven pundits in Varanasi and Mathura are performing a special *anushthan* (ritual) that will continue round-the-clock for seven days at the Vishwanath temple.

At the end of the *anushthan*, an elaborate *rudra abhishek* will take place and that, everyone hopes, will put an end to all the troubles that are presently plaguing UP chief minister Mulayam Singh Yadav.

BJP leader Rajnath Singh's supporters last month offered a special *puja* at the Shakumbhari Devi temple in Saharanpur as soon as his name began doing the rounds for the BJP president's post. The prayers were promptly answered and Mr Singh became party chief. His supporters are now preparing for *shringar* as a thanksgiving gesture.

Samajwadi MP Amar Singh's supporters, on the other hand, are also planning a *maha mrityunjaya jaap* to help their leader overcome his present crop of problems, including those of the telephone tapping. Trouble, they say, takes you closer to God and this, perhaps, holds truer for politicians than for ordinary mortals.

When trouble knocks—and it seems to be doing so rather regularly these days—our *netas* rush to temples, order elaborate *pujas* and adorn (the temple with) precious gems. Their priests, too, revel in the political attention that is lavished on them and happily publicise the 'help' that they have extended to these *netas*—albeit, on condition of 'secrecy'. Almost every Uttar Pradesh politician, including those who belong to the secular variety, are known to hold secret *pujas* and elaborate rituals to ensure their political well-being. Some even believe in rituals that are designed to 'harm' their rivals but these are often conducted in utmost secrecy.

A leading priest of Varanasi who has performed *pujas* for the leading politicians at one time or the other, says rather candidly, "We are busy round the year because one astrologer or the other prescribes *pujas* for the leaders which we are asked

* *Deccan Chronicle*, Hyderabad, 24th January 2006

to perform. In most cases, the leaders send their supporters with a request for the *puja* and also express their inability to remain present during the *puja* because they do not wish to be seen praying in public. We usually keep a photo of the leader and perform the *puja*. Moreover, it does not make a difference to us as long as we get our *dakshina*."

A priest in Mathura discloses that a special *anushthan* is currently underway in Vrindavan for Mr Mulayam Singh Yadav. "The *anushthan* began on Makar Sankranti and will continue for seven days. Mr Yadav is passing through a bad phase and his stars will give him a tough time in court cases. The *anushthan*, which is costing Rs.51,000, will give him some relief and it is one of his ministers who is organizing it," says a local priest.

The *dakshina*, incidentally, often extends beyond the temple, and priests who perform *pujas* for politicians end up getting huge favours from their "clients".

A *tantrik* in Lucknow was gifted a plot and a car by a former chief minister when he helped the former during a political crisis. A priest in Mathura recently got his two brothers a government job after he performed *puja* for an influential minister. Other priests in Varanasi and Kanpur have earned lucrative building contracts for their kin after they came into contact with their political clients.

While Mr Yadav's supporters are heading for temple towns like Varanasi and Mathura, former chief minister Kalyan Singh's faith in the Vindhyavasini temple in Mirzapur is well-known.

Though Mr Kalyan Singh, as UP chief minister in 1991, had declared that he had a direct 'hotline' with Lord Ram, and could communicate with Him whenever he wanted, his loyalties now seem to have shifted to Vindhyavasini Devi. Mr Singh, along with his close associate, Ms Kusum Rai, is a regular visitor to the Vindhyavasini temple and participates in elaborate rituals at the temple whenever he faces political trouble—which is rather often these days. During Navratri, the former chief minister makes it a point to visit the Vindhyavasini temple and the local priests happily relate tales of the VIP's faith in the goddess.

Another senior BJP leader who is known for his faith in the Almighty is Mr Kalraj Misra who makes a beeline for Varanasi at regular intervals and is known to participate in *rudra abhishek* at the Vishwanath temple.

Ms. Uma Bharti, like most of the other BJP leaders, makes it a point to visit the makeshift Ram Lalla temple in Ayodhya whenever she is touring UP. This not only gives her an opportunity to reiterate her commitment to building a Ram temple but also allows local BJP activists to chant the time-tested 'Jai Shri Ram' slogan.

However, ever since Ayodhya in general, and the Ram temple in particular, acquired political significance, most of the secular leaders, including those from the Congress and the Samajwadi Party, have shied away from the temple town. As a veteran Congress leader says, "Visiting Ayodhya has become almost synonymous with supporting the construction of the Ram temple. Though I used to visit Hanuman Garhi in Ayodhya regularly with my family till about a decade ago, I have stopped going there now…".

* * *

MOCK FUNERAL TO DRIVE AWAY BAD LUCK?*

In a bizarre incident, a family performed the funeral rites for their three children, who are alive, because of superstition. A crow had struck them while they were playing in Ramreddypalli village of Mustabad Mandal on Sunday and this was considered a bad omen.

Reports said the children Anitha (10), Saikumar (8) and Shivkumar (2), were playing with chicks in front of their house, when a crow trying to pick up a chick, struck against them. The parents, who were watching, immediately instructed the children to sleep like dead persons and informed their relatives that the children had drowned in a well.

Relatives from far-flung places including Hyderabad thronged the village and wailed in sorrow. Later, the parents told them the real story.

* * *

* *The Hindu*, Karimnagar, 17th January 2006

"RUHANIYAT IS LIKE FIRE": PIR HAJI KASHANI BABA SPEAKS TO SHIVAM VIJ*

A family was driving towards Noida when their son wanted to take a leak. He got out of the car, turned away from the road, faced the Yamuna and relieved himself. A few minutes later, he fell ill and started behaving abnormally. His horrified family tried everything they could; finally they took him to Syed Mansoor Nizami, alias Pir Haji Kashani Baba, in Nizamuddin, who instantly divined that a spirit had possessed the boy.

The Baba sat the boy down before him and addressed the spirit. "O Spirit, why did you enter the body of this boy?"

When a spirit enters a person, it is the spirit who speaks, and this one said the boy had urinated upon it on the banks of the Yamuna.

"It is perfectly understandable", said the Baba. "Anyone you pee on would be angry at you."

Pir Haji Kashani Baba then proceeded to exorcise the spirit with the help of a broom made of peacock feathers. The boy was soon restored to his senses.

This, at least, is the Baba's version of events. He has been exorcising spirits for decades and, according to him, the most common reason why they possess human beings is the scent of perfume or the flowers that women wear.

"When women go on family picnics to India Gate, they don't realize that the trees in a place like that are at least 200 years old", he says, "Evil spirits hang from the branches of old trees and are attracted by perfume."

Committing suicide amounts to deciding one's fate and not letting Allah do so, he says, and thus the spirits of those who kill themselves never reach either heaven or hell, and are condemned to linger on earth forever. These are the evil spirits, the *shaitani atma,* that possess people.

Pir Haji Kashani Baba can change your business fortunes, get you married, prevent your divorce, make you conceive (and

* *Tehelka*, 24th February 2007

grant you a boy at that), save your children from the evil eye, help them perform better in school, and bless them for life. And exorcise spirits. He can do this because he has what most of us don't—the power of *ruhaniyat*. "*Ruhaniyat* is the power of Allah," he says, "It comes from the will of Allah. Allah's wish is that you follow him, perform *namaz* regularly, read the Quran Sharief. If your karma is good, Allah may bless you with *ruhaniyat*."

Once acquired, though, *ruhaniyat* is like a ball of fire. If not used carefully, it can ruin its owner. "*Ruhaniyat* is to be felt, *ruhaniyat* is *aatishi, ruhaniyat* is the power of Allah. Should Allah want, he can turn day into night and night into day."

"I am more like a doctor or engineer, you know, who uses specialized knowledge," he says, adding, however, that he doesn't have much respect for scientists. "They don't believe in Allah."

There are many Sufis in Nizamuddin who claim to have such powers and who distribute amulets and pray for you. But Pir Haji Kashani Baba is different-for, unlike many others, he has a two-story establishment all to himself. Its hoardings apart, it looks rather ordinary, much like any of the other shops cramping the mausoleum of Hazrat Khwaja Nizamuddin Auliya, the famous 12th century Sufi saint. The Baba's shop, like his visiting card, advertises: "Spiritual Treatment for the solution of any kind of problem, Business and Marriage Purpose, Effect of Bad Air, *Jadu Tona*". He also has a website (www.sufisainthazratnizamuddinaulia.com) which is currently down, because he's planning renovations with more extensive information about *ruhaniyat*. Requests are received by post or e-mail, but he prefers most to be faxed. A letter lying near him lists the names of all the members of a joint family that is consulting him, with flowchart-like lines indicating who is doing *jadu tona* on whom. Since his fame spread across the world, all sorts of people write to him: white, black and brown. A request from the Netherlands includes a family photograph. "This is my family," it says on the back. "Help me save it." The divorce successfully prevented, the entire family flew down to

Delhi to pay obeisance at the Nizamuddin Dargah.

For all his emphasis on following the path of Allah, the Baba is quick to say that all religions are equal. "I serve humanity, not religion." The people who come to him are largely Hindu and Sikh; when people's miseries are not alleviated by their own religions, they step away from them and into the plural domain of the Sufi tradition. Those complaining from 'Effect of Bad Air' or *Jadu Tona* may also be suffering from mental disorders.

On the ground floor at the Baba's is the waiting area. Remove your shoes outside and wait. If there is no attendant, a seemingly disembodied voice will command you to come upstairs. You obey and enter a plush, air-conditioned room; you may not notice the CCTV camera with which the Baba anticipates his visitors or the microphone over which he announces whose turn is next. There are several phones, mobiles, a fax machine, and if you don't drink tea, there's always Pepsi. But the traditions of centuries are in no conflict with modernity: "The outward appearance is immaterial. *Ruhaniyat* is in the heart and the mind."

A prominent signboard in the Baba's office says, "Ask only one question in one time." He clarifies that once people have requested him to solve the problem they came with, they start asking solutions for irrelevant matters like stomach aches or servants who have run away."

So how much does he charge? He points to the donation box. "I don't need people or their money. They need me," he says grandly. "Fakirs are the *shehenshahs* of Time."

The one thing *ruhaniyat* can't do is stop wars "because they happen of Allah's will. They happen because man has been running after the *shaitaan.*" The one thing he won't use *ruhaniyat* for is winning you an election. "A politician himself once told me that a politician is born after 100 politicians die." It was to keep away from politics that he refused to participate in peace processions in the aftermath of the Babri Masjid demolition. And he says he won't entertain requests to use *ruhaniyat* to hurt another person.

Often you may not find him in his Nizamuddin office, for he travels the world over. People send him air tickets and sponsor his stay. All he does is use the power of *ruhaniyat* to fulfil their wishes. Lest you think this claim is a lie, he shows you his passport, five old ones stuck together. In the last month-and-a-half, he's visited 11 countries. Don't his followers in Delhi suffer when he is away? "What to do, Allah takes me all over the world." This includes his office in Texas, a photograph of which adorns his booklet about the Nizamuddin tradition and his place in it.

He draws his lineage from Hazrat Nizamuddin Auliya's sister, and he is also the president of the Chishtia Nizamia Mission, which takes care of the shrine and spreads the saint's message to the world.

The Baba is the 31st in his line and his interest in *ruhaniyat* was sparked when, at age 10, he saw his father save a man's life by blessing him. Pir Haji Kashani Baba's two elder sons, however, run a CD manufacturing business, and his third is training to be a software engineer. Doesn't he feel sad that the tradition of *ruhaniyat* in his family could come to an end? "Why? Just as people come to Hazrat Nizamuddin's dargah with their wishes, so they will come to mine."

* * *

What have all these stories—and innumerable others like them that could be told, that take place everywhere and involve every section of our society—have in common? They relate to a startling lack of scientific temper in our people: be it belief in miracles or miracle-men, astrology or homoeopathy.

Indeed, one of the main reasons for our not doing much better as a nation—one of the reasons for many of our ills—is the lack of scientific temper amongst our people, and this is in spite of our being one of the ten scientifically and technologically most advanced nations in the world.

From Chief Ministers attempting to bring luck and stability by changing the directions of their tables in their offices or relocating entrances of Secretariats in accordance with *vaastu shastra,* to thousands of asthma patients flocking to Hyderabad on the 8th of June every year to swallow the famous fish medicine considered to be a divine cure, scientific temper is the first to be thrown out of the window by the rich and the poor, the powerful and the lowly. And scientific temper does not get butchered by lay people alone, scientists do so with equal glee. Thus we had, in this decade, the replica of a satellite taken to Tirupati for Balaji's blessings by the scientists of Indian Space Research Organization before its launch. Not only auspicious dates for marriages or house warming ceremonies but also avoiding *rahu kalam* for meetings and journeys are practices that continue to be keenly observed even by a large section of the scientific fraternity. Several of our well-known scientists (one of whom was even considered for the Nobel Prize) have been ardent followers of godmen.

Thus lack of scientific temper pervades every section

of society and raises its ugly head in every human activity in our country. Perhaps, only the poorest of the poor such as street-dwellers who have spent their entire lives on pavements under the open sky through scorching heat, biting cold and pouring clouds, and have never had illusions of having received any favour from any god, harbour no superstition. In fact, it may be that as we go up the hierarchy of affluence in our society, the extent of irrationality and belief increases.

Yet, it would be wrong to conclude that our country is the only one amongst the community of nations where scientific temper does not prevail. One only has to look at the raging war for space in school curriculum in the United States, between "intelligent design" and Darwinian evolution. It is indeed ironical that the pressure to teach creationism in schools has been growing in the country where the largest amount of work on the human genome has been done, which has proved that humans are 99 per cent chimpanzees in their genome, thereby supporting the fact that man has evolved from lower forms of life. Another recent example of such blatant obscurantism in the West was the cannonisation of Mother Teresa, which required proof of miracles having been performed by the late Mother Teresa. And two miracles that never occurred were invented amidst wide-spread criticism from rationalists all over the world, to confer sainthood on her.

The lack of scientific temper is, therefore, as much a cause of problems around the rest of the world, as it is in India. (It is surely not the *only* cause but an important one.)

In fact, we ought to do better than anyone else in respect of adherence to scientific temper; for we have an extraordinary advantage. The term, 'scientific temper', was coined in India, and we are the only country in the world which has it as a duty of its citizens.

This book is a story of development as well as negation

of scientific temper in India as seen through our eyes. (The two authors, Pushpa M Bhargava and Chandana Chakrabarti, would be referred to as PMB and CC, respectively.) It is largely a collection of our articles, or published material regarding incidents and happenings in the country that related to scientific temper in which at least one of us was involved. Some material has been repeated, in content, in more than one chapter. We have not eliminated this repetition totally while editing the articles, to ensure the independent readability of each chapter.

The book is not an exhaustive compilation of material on scientific temper. There are many individuals and organizations in the country, not mentioned in this book, who have contributed substantially—perhaps much more than we have—towards disseminating scientific temper; each one of them could narrate their own story, may be much more exciting than ours. This book is, in a way, our homage to them.

Chandana Chakrabarti Pushpa M Bhargava

II

THE STORY OF SCIENTIFIC TEMPER IN INDIA IN A NUTSHELL

Nehru, The Scientific Temper, and the Association of Scientific Workers of India

Jawaharlal Nehru coined the term, 'scientific temper', in his book, *The Discovery of India*, in 1946. He was also the first of our leading politicians who talked about the conflict between science and religious dogma: *not* the basic values which all religions preach and which are, for all practical purposes, identical, *but* the dogma that gives a religion its identity; different religions have a totally different set of dogmas which are entirely a matter of belief and for which no rational or reliable evidence exists.

Nehru became the first President of the Association of Scientific Workers of India (ASWI), a trade union of scientists who had at least a B.Sc. degree or equivalent, founded in the late 1940s. PMB had the privilege of being a member of the first Executive Committee of the ASWI, one of the main objectives of which was to develop scientific temper. Nehru was the first to recognize the role of scientific temper in the development and progress of the country.

Unfortunately, over the decades, the ASWI, which had at one time—say, the early 1950s—most of the leaders of the scientific community as members, has declined into virtual non-existence. For example, the Hyderabad branch of which PMB was the Secretary was, between 1950 and 1953, very active with nearly 500 members, including

Dr M Channa Reddy who later became the Chief Minister of Andhra Pradesh. Dr Channa Reddy himself later succumbed to superstition as evidenced by the fact that he carried a lucky wand in his hand given by a godman, so that power would continue to remain in his hands! The Hyderabad branch of ASWI has been extinct for sometime.

Nehru's penchant for scientific temper is described in Chapter III; it is an excerpt from the article, "A reassessment of the contribution of Jawaharlal Nehru to science", in the book, *Nehru Revisited*, edited by M V Kamath and published by Nehru Centre, Mumbai.

Society for the Promotion of Scientific Temper

In 1963, with the falling credibility of ASWI, Satish Dhawan who later became one of our foremost space scientists and the Chairman of Indian Space Research Organization, Abdur Rahman, the historian of science, and PMB, felt that a national society set up exclusively for promotion of scientific temper could be a social asset. Thus they prepared a statement which was published in 1964 in *Seminar*, issue no.55, pages 10-11. This statement (reproduced in Chapter IV) was used to launch the Society for the Promotion of Scientific Temper at the occasion of an international symposium on nucleic acids held in the then Regional Research Laboratory (today, the Indian Institute of Chemical Technology) at Hyderabad in January 1964. The address of the Society was Post Box no.237, New Delhi-1!

The above-mentioned symposium was India's first major international meeting in molecular biology and related fields in India; it was attended by most of the leading molecular biologists of that time from across the world, including the Nobel Prize-winning co-discoverer of the structure of DNA, Francis Crick. (This meeting is now a part of the history of biology as several major discoveries in biology were reported for the first time at this meeting.)

Many international participants in the above symposium, including Francis Crick, along with many well-known and progressive Indians such as S. Husain Zaheer (the then Director-General of CSIR, Government of India), Abdur Rahman, Satish Dhawan, Mohit Sen and Maqdoom Mohiuddin (well-known leaders of the Communist movement in India), were present at the launch of the Society for the Promotion of the Scientific Temper; they all strongly supported the statement on scientific temper.

Membership to this organization required that an applicant signs the following declaration:

> I believe that knowledge can be acquired only through human endeavour and not through revelation, and that all problems can and must be faced in terms of man's moral and intellectual resources without invoking supernatural powers.

This seemingly innocuous requirement, unfortunately, turned out to be the undoing of the fledgling Society. All those who were involved in the setting up of the Society and those who actively worked for it later (which included many highly reputed academicians), had assumed axiomatically that scientists of the country would have no hesitation in signing this declaration which merely reflected the basic spirit of scientific temper. By definition, scientific temper, denies that any "external" agency which would be outside the purview of science can have any role in solving any problem—individual or collective.

All those concerned were, therefore, taken aback when scientist after scientist across the country refused to sign the declaration. It demonstrated to their great chagrin and disillusionment the extreme lack of scientific temper in the scientific community itself—a situation that largely continues.

We believe that it is this situation that is responsible in a major way for our many failures in science, and this is not to underrate our successes in science and technology which we have documented with pride in our book, *The Saga of Indian Science Since Independence: In a Nutshell* (Universities Press, 2003). It did not take us long to realize that the situation was in total contrast with that prevailing then in the top scientific community of the "developed" (scientifically and technologically advanced) countries around the world.

The Society for the Promotion of Scientific Temper died a natural death: this chapter on development of scientific temper in the country was closed but many lessons were learnt from it, one of them being that scientific temper was an important ingredient of any recipe for not only social and economic but also scientific and technological advancement of our country.

Other Articles Between 1964-1980

Between 1964 and 1980, a number of articles relating to scientific temper were published by PMB and his colleagues. Some of them are listed below and reproduced in Chapters V-XII.

Chapter V:	The scientific outlook
Chapter VI:	Indian society and the scientific temper
Chapter VII:	Obscurantism and academics
Chapter VIII:	The scientific temper
Chapter IX:	The phenomenon of Maharishi Mahesh Yogi
Chapter X:	The method of science and the value system implicit in it
Chapter XI:	Does science refute religion?
Chapter XII:	Science and dogma

During this period, many proponents of scientific temper—individuals and organizations—emerged in the country that made a vital social contribution. Examples would be Kerala Shastra Sahitya Parishad that played a major role in achieving total literacy in Kerala, the Bangalore Science Forum, Paschim Banga Vigyan Manch, Marathi Vigyan Parishad, and Jana Vignana Vedika of Andhra Pradesh. Their rational and often successful efforts would need separate books to describe and discuss.

The Method of Science Exhibition

In 1975, PMB was asked by Dr Rais Ahmed, the then Director of the National Council for Education Research and Training, to prepare a national exhibition on the method of science. This exhibition was prepared in Hyderabad at the then Regional Research Laboratory (now the Indian Institute of Chemical Technology), between 1975 and 1976 with financial and other support from a large number of organizations in India and abroad. The exhibition was set up in Bal Bhavan in Delhi in a 5,000 sq.ft. independent building (called the Polish pavilion) during January-March 1977. It was to be inaugurated by the then Prime Minister, Indira Gandhi. She, however, lost the election towards the end of March 1977. The travails of the exhibition then began.

The Morarji Desai Government that followed, found the exhibition undesirable in many ways, particularly as it emphasized questioning; no totalitarian or sectarian regime like the one we had at that time, likes to encourage questioning. For this and other equally flimsy reasons that pertained to certain contents of the Exhibition, the above Government arranged for it to be surreptitiously dismantled and removed ('stolen', as it was universally felt) in the first week of August 1988. This clandestine operation was done secretly at night, without anyone knowing anything about it. This incident raised much public hue and cry—nationally and internationally. It led to several enquiries, one of them initiated by the PM's office, and to a court case. The stolen exhibition was finally located, and purchased in a badly damaged condition by the Andhra Pradesh Government. It was then redone and launched in Hyderabad amidst much fanfare, where it became a major academic attraction. However, after a while, it lost the patronage of the Government and once again fell into disarray. It was then transferred with the permission of its creator, PMB, to Birla Science Centre in Hyderabad, where it was never exhibited properly and eventually, after a few years, it went through a second process of destruction on account of, perhaps, the same reasons that led to its first virtual destruction. Interestingly, the only objects from the exhibition that survived were some priceless paintings by Laxma Goud and Surya Prakash. These paintings which were an integral part of the exhibition were surreptitiously added to the personal collection of Smt Nirmala Birla. We discovered this only when they were exhibited with much élan in an art exhibition of the personal collection of Smt Birla at the same venue where the exhibition was initially set up. Perhaps, in course of time, other priceless memorabilia and paintings that were an integral part of the exhibition and a labour of love by some of India's best known painters would find their way into the same personal art collection, for art has a price whereas scientific temper asks for a price of the exploiters and the exploited!

The story of this much publicised exhibition is given in Chapter XIII in PMB's words. This is an excerpt from the 920-page book titled, *Vandalisation of a Work of Art and Science*, edited and published by B. Premanand for Geedee Medical Aids in 2005, in which all the material pertaining to the Exhibition including correspondence, press reports, reports of the enquiry and the court case, has been put together. Chapters XIV and XV reproduce two articles appearing on this Exhibition in two of the world's best-known scientific journals, *Science* and *Nature*.

The Statement on Scientific Temper

In 1979, PMB was invited to a conference at the Tata Institute for Social Sciences in Bombay where he met for the first time, Bakul Patel, who was then involved in setting up the Nehru Centre in Bombay with her husband, Rajni Patel, one of the best-known criminal lawyers of the country. (Bakul Patel later became the Sheriff of Bombay.) Their common commitment to scientific temper and rationalism, led to a lasting friendship. Together, in October 1980, they held a meeting on scientific temper in Coonoor, which was presided by P N Haksar, one of the most distinguished civil servants our country has ever had.

The above meeting led to "A statement on scientific temper" which was signed by some of the best-known intellectuals in the country, in addition to the participants of the meeting; many others wrote about their support to the statement—which was widely publicized and debated in the country—later, after publication of the statement. It is reproduced in Chapter XVI of the present book. Some of the comments on this statement have been put together in the book, *Science and Sensibility* by K V Subbaram, published by Manthan Publications, Rohtak, in 1989.

Amongst the many commentaries on the Statement on Scientific Temper, perhaps the most noteworthy was that of the late Swami Ranganathananda who later became the head of the Ramakrishna Mission. This commentary led to the Nehru Centre organizing a debate in Mumbai about a year later, between a group of 15 persons chosen and headed by Swami Ranganathananda that felt that there was an alternative to the Statement on Scientific Temper, and another group of 15 selected by the Nehru Centre. The group of 30 unanimously elected Mr P N Haskar to chair the debate which was spread over three days, and followed all the norms of a healthy, strictly academically-oriented debate. All the seats open to the public in the hall in the 'Discovery of India' Pavilion of the Nehru Centre (that was still under construction at that time) were booked in advance; the public was invited to take part in the debate in the afternoons. Although at the end of the debate, the consensus was that the Nehru Centre team had won but that was not so important. What was important was the high quality of the debate and the exemplary manner in which it was conducted—perhaps a fitting ode to Indian culture and to scientific temper. The debate was widely covered by the press.

An article by PMB, "Why (the above) statement on scientific temper?", is reproduced in Chapter XVII.

The Scientific Temper and Method of Science in History in India

Indian successes in science and technology from the time of Harappa and Mohenjadaro (that is, some 5,000 years ago) to the end of the 19th century are well-known and well documented. India was, during the above period, a leader in many areas such as mathematics, astronomy, medicine and surgery. For example, no other country in the world had a mathematician like Bhaskaracharya, whose equally talented daughter, Leelavati, had a flair for numbers and could solve numerous mathematical riddles given by her father in Sanskrit *Shlokas* (verses), such as the following:

- A herd of elephants. A half and a third of that half went into a cave. One sixth and a seventh of that were quenching their thirst on the river and one eighth and one ninth of that eighth were enjoying themselves in a lake full of lotuses. The chief of the herd and his three beloveds were playfully engaged in a love game. How many elephants were in the herd?
- From a heap of lotuses, one third were offered to Lord Shankara, a fifth to Vishnu, one-sixth to the sun god, and one fourth to the goddess. The remainder of six lotuses

was laid at the feet of the Guru. Tell me the total count.
[The above two slokas show how Bhaskaracharya brought life and excitement into the dry sums of fractions, ratios and proportions.]

- A lake inhabited by a number of *Chakravaka* and *Kraunch* birds. There stands a lotus half-a-hand above the water level. A light breeze displaces it to submerge two hands away. Tell me quickly the depth of water.
[This very picturesque scene is an ideal setting to demonstrate what is known today as Pythagoras theorem. Visualize the submerged lotus and its original state before the breeze does its act, and the familiar right-angled triangle would become evident.]

- How many different types of idols of Lord Shiva would you get if his ten hands were to hold ten different weapons—arrow, bow and snake et al.—every time changing their configuration. Likewise, how many idols of Lord Vishnu would be possible with him holding four such things?
[*The concept of "permutations"—arrangements of objects in a distinctly identified manner—is explained in this shloka by visualizing different images of Shiva and Vishnu, each distinguished from the other by the configuration of weapons.]*

- Bees equal to the square root of one-half from a swarm flew towards the *Malati* trees. Eight-ninth of the swarm also followed them. Out of the two bees remaining, one was lured by the sweet smelling lotus and got entangled. His outcry was responded by his mate. Tell me, dear, the number of bees in the swarm.
[*If one thought that solving a quadratic equation was quite boring, this swarm of bees would dispel that notion.]*

- There is a pole, nine hands tall. At its feet is a snake hole. A peacock is perched on top of the pole. A snake approaches the hold from a distance thrice that of the height of the pole. The peacock sights the snake, swoops down diagonally at the same speed and grabs it at a distance from the hold. Will you tell me quickly what the distance is?

[*Like the submerged lotus, this one too demonstrates the use of right-angled triangles.*]

- Four streams feed a well. Each one on its own fills up the well in a half, a third, a fourth and one full day. If all four flow simultaneously, how much time would it take to fill up the well ?
- There was a lake full of dainty lotuses and a flock of swans. Once when the sky was overcast with clouds, ten times the square root of the total number in the flock flew towards Lake Manas. One-eighth of the flock hurried to the woods full of hibiscus. Left behind in the lake were only three lovelorn pairs. Tell me my little girl, the number of swans in the flock.
- A man gifts his beloved a number of gems to adorn her with ornaments. One-eighth of these beautify the parting of her hair, three-seventh of the remaining go into a necklace long enough to rest in the cleavage of her breasts, half of the remaining gems she wears in the form of armlets. Out of the gems left after that, three fourths make a waistband with jingling bells. Sixteen gems were left over with which she bedecked the plaits of her hair. Tell me quickly the total number of gems.
- An oil-lamp and a cone are standing three hands apart. The lamp is three-and-a-half hands tall; the height of the cone is half-a-hand. Tell me how far the shadow of the cone would fall.

  [*This is one of the more ingenious of Bhaskaracharya's real-life-oriented problems. Visualize the oil-lamp and its rays falling on the cone standing some distance apart. It is easy to see the usage of rule governing the ratios of sides in similar triangles.*]

- In an intense foreplay the pearl-necklace of a woman falls apart. One-third drop down on the floor and one-fifth slip under the bed. She manages to pick up one-sixth and the man lays his hands on one-tenth. Finally only six pearls are seen locked up in the thread. Tell me how many pearls the necklace was made of?

These riddles have been exquisitely choreographed in an Odissi dance ballet, "Leelavathi", by Jhelum Paranjpye.

A legitimate question to ask, therefore, would be: what was the status of scientific temper, and of the method of science an understanding of which comes instinctively with scientific temper, in ancient and mediaeval India. Chapter XVIII reproduces an article by us, "The scientific temper and the scientific method in science in India through history, with special reference to biology".

Some Other Articles (1989-1990)

Finally, we reproduce here the following articles by one or both of us that relate to scientific temper:

Chapter XIX:	Secularism and the scientific temper
Chapter XX:	Modernity and the scientific temper
Chapter XXI:	The seven deadly sins of the clergy

The last article is included as the clergy have been, we believe, the primary stumbling block in the spread of scientific temper which, unlike the clergy's religions (be they Christianity or Islam, Hinduism, Sikhism or Judaism) that divide people, unites them.

III

NEHRU AND THE SCIENTIFIC TEMPER*

P M Bhargava

Nehru was acutely aware of the changes that had come about by the end of World War II and the Indian independence, in respect of rights and responsibilities of scientists. He also was fully cognizant of the hold of irrationality, religious dogma and superstition amongst our people, including the scientists. He, therefore, rightly realized that for the scientists to give their best to science and to society, they must have scientific temper, and he expected them to be the main instrument for disseminating the temper of science amongst the people of our country. He, in fact, coined the term, "scientific temper", in his book, *The Discovery of India*, in 1946. He said:

> The applications of science are inevitable and unavoidable for all countries and people today. But something more than its application is necessary. It is the scientific approach, the adventurous and yet critical temper of science, the search for truth and new knowledge, the refusal to accept anything without testing

* This is an excerpt from the chapter: "A reassessment of the contribution of Jawaharlal Nehru to science" from the book, *Nehru Revisited*, edited by M V Kamath and published by Nehru Centre, Discovery of India Building, Dr Annie Besant Road, Worli, Mumbai 400018, 2003, pp.42-48

> and trial, the capacity to change previous conclusions in the face of new evidence, the reliance on observed fact and not on pre-conceived theory, the hard discipline of the mind—all this is necessary, not merely for the application of science but for life itself and the solution of its many problems. Too many scientists today, who swear by science, forget all about it outside their particular spheres. The scientific approach and temper are, or should be, a way of life, a process of thinking, a method of acting and associating with our fellowmen. That is a large order and undoubtedly very few of us, if any at all, can function in this way with even partial success. But this criticism applies in equal or even greater measure to all the injunctions, which philosophy and religion have laid upon us. The scientific temper points out the way along which man should travel. It is the temper of a free man. We live in a scientific age, so we are told, but there is little evidence of this temper in the people anywhere or even in their leaders.

His criticism of religion, as identified by religious dogma, was fair and unemotional. For example, in his book, *The Discovery of India*, he said:

> Religions have helped greatly in the development of humanity. They have laid down values and standards and have pointed out principles for the guidance of human life. But with all the good they have done, they have also tried to imprison truth in set forms and dogmas, and encouraged ceremonials and practices which soon lose all their original meaning and become mere routine. While impressing upon man the awe and mystery of the unknown that surrounds him on all sides, they have discouraged him from trying to understand not only the unknown but what might come in the way of social effort. Instead of encouraging curiosity and thought, they have preached a philosophy of submission to nature, to the

> established church, to the prevailing social order, and to everything that is. The belief in a supernatural agency which ordains everything has led to certain irresponsibility on the social plane, and sentimentality has taken the place of reasoned thought and inquiry. Religion, though it has undoubtedly brought comfort to innumerable human beings and stabilized society by its values, has checked the tendency to change and progress inherent in human society.

And again in the same book while comparing the scientific method with the method of religion, he said:

> Very different is the method of religion. Concerned as it is principally with the regions beyond the reach of objective inquiry, it relies on emotion and intuition. And then it applies this method to everything in life, even to those things which are capable of intellectual inquiry and observation. Organized religion, allying itself to theology and often more concerned with its vested interests than with things of the spirit, encourages a temper which is the very opposite to that of science. It produces narrowness and intolerance, credulity and superstition, emotionalism and irrationalism. It tends to close and limit the mind of man, and to produce a temper of a dependent, unfree person.
>
> Even if God did not exist, it would be necessary to invent Him, so Voltaire said—*'si Dieu n'existait pas, il faudrait l'inventer.* Perhaps that is true, and indeed the mind of man has always been trying to fashion some such mental image or conception which grew with the mind's growth. But there is something also in the reverse proposition: even if God exists, it may be desirable not to look up to Him or to rely upon Him. Too much dependence on supernatural factors may lead, and has often led, to a loss of self-reliance in man and to a blunting of his capacity and creative ability.

> As knowledge advances, the domain of religion, in the narrow sense of the word, shrinks. The more we understand life and nature, the less we look for supernatural causes. Whatever we can understand and control ceases to be a mystery. The processes of agriculture, the food we eat, the clothes we wear, our social relations, were all at one time under the domain of religion and its high priests. Gradually, they have passed out of its control and become subjects for scientific study. Yet much of this is still powerfully affected by religious beliefs and the superstitions that accompany them. The final mysteries still remain far beyond the reach of the human mind and are likely to continue to remain so. But so many of life's mysteries are capable of and await solution, that an obsession with the final mystery seems hardly necessary or justified. Life still offers not only the loveliness of the world but also the exciting adventure of fresh and never-ceasing discoveries, of new panoramas opening out and new ways of living, adding to its fullness and ever making it richer and more complete. It is therefore with the temper and approach of science, allied to philosophy, and with reverence for all that lies beyond, that we must face life.

The above thoughts are echoed in almost all of Nehru's speeches in one context or another and, in a way, defined his perception of scientific temper. It is relevant to state here that it was the Nehru Centre at Bombay which issued the famous Statement on Scientific Temper in 1981, with the formulation of which I was intimately associated, along with Bakul Patel, Rajni Patel, Raja Ramanna and P N Haksar.

It was Nehru's concern for rights and responsibilities of the scientists that made him accept the first Presidentship of the Association of Scientific Workers of India (ASWI) at its Annual General Meeting on January 8, 1947. ASWI was registered as a trade union and affiliated to the World

Federation of Scientific Workers with which many famous scientists, such as Cecil Powell, the Nobel Prize winning physicist from UK, Ivan Malek, the well-known microbiologist from Czechoslovakia, and Frederic Joliot-Curie, the Nobel Prize winning son-in-law of Marie Curie, were also associated. Nehru's, indeed, may be the only case where the executive head of a State has concurrently headed a trade union.

While accepting the Presidentship of the ASWI on January 7, 1947, he said:

> I do not want that the man who receives honours should go without any money. I hope that in the new set up that we are likely to have, money at any rate will not have too much power or honour or glory attached to it and that honour will go to service and learning.

This was a superb though cryptic statement of his perception of the rights and responsibilities of scientists.

A week later, on January 14, 1947, he gave a call to scientists all over the country to become members of the ASWI which became a powerful organization in the early 1950s with a large and distinguished membership around the country. I recall that in the early 1950s, the Hyderabad Branch of ASWI, of which I was the Secretary, had more than 400 active members including Dr M Channa Reddy, who later on became the Chief Minister of Andhra Pradesh.

On January 2, 1956, Nehru addressed the Annual General Meeting of the ASWI at Agra and made two important points:

> Unfortunately, the Universities were still not in a position to promote the development of research for want of funds," and he doubted, "if trade unionism, which was unavoidable for industry, would help scientific workers.

Jawaharlal Nehru was the President of ASWI only in 1947-48 but continued to believe in the strength and value of the organization for years afterwards. He thus attended the tenth meeting of the Council of ASWI at Madras on January 2, 1958, when Major General S S Sokhey was the President of ASWI. At that occasion Nehru again said that the "example of industry does not fully apply to scientific workers", and talked about both the rights and the responsibilities of scientific workers. He specially talked about the professional and social accountability of scientists. (He assumed that scientists will not commit financial impropriety!)

Perhaps, the most effective demonstration of his belief in the scientific method—a prerequisite for a scientific temper—was his setting up the Planning Commission; he was aware that the concept of planning national growth and development in the Soviet Union and France had been a consequence of the application of the scientific method to social problems. He found in P C Mahalanobis, someone who had the right background for using this method appropriately in planning.

In the 1955 session of the Indian Science Congress, Nehru said, "I myself am not bound by dogmas and am always prepared to admit my mistakes and to rectify them. I believe that such an approach is nearer to what may be called the scientific approach, and in that sense I consider myself having a scientific temperament, although I cannot claim to be a scientist."

To be able to say, "I don't know", or that "I made a mistake", is an important attribute of the scientific method.

Nehru had an intrinsic respect for scientists and felt that they had the right to hold opinions which were their own. Let me illustrate these two points with one incident each.

In April 1957, the Ramanujam Institute of Mathematics

in Madras founded by Sir Alagappa Chettiar, was in trouble. S Chandrasekhar—the astrophysicist who later on won a Nobel Prize—wrote to Nehru about this. Nehru took immediate action and the Institute is very much alive today.

In the early 1950s, Dr S Husain Zaheer, the Director of the then Central Laboratories for Scientific and Industrial Research under the Government of the State of Hyderabad (which laboratory was taken over by the CSIR later and renamed first as the Regional Research Laboratory, Hyderabad, and subsequently the Indian Institute of Chemical Technology), received a note from the Government of India asking him to terminate the services of one of his Employees. Dr Zaheer refused to do so as he felt that such a termination of services would be violating the rights of the individual if no reason for doing so was given to him. Finally, the reason came from Delhi: that the scientist was an active member of the Communist Party of India before Independence, when he was a student in the erstwhile-undivided Punjab. This colleague of mine then had to be suspended but we all, including Dr Zaheer and people like Dr K S Krishnan, fought his case. When it was finally brought to the notice of Jawaharlal Nehru, he called our suspended colleague, gave him the fullest attention and issued orders for his reinstatement and payment of all his salary arrears.

It was the Nehru legacy of scientific temper that encouraged several of us, specially Prof. Nurul Hasan, to persuade Mrs Indira Gandhi to include the following provision in the 42nd amendment to the Constitution of India:

> It shall be the duty of every citizen of India to develop a scientific temper, humanism and the spirit of enquiry and reform.

IV

SOCIETY FOR THE PROMOTION OF SCIENTIFIC TEMPER*

POST BOX NO.237, NEW DELHI - 1

Dear Friend,

This letter cordially invites you to become a member of the Society for the Promotion of Scientific Temper. The objectives of the Society are given in the enclosed folder.

We believe that promotion of a scientific temper is the need of the hour and can play a major role in the moral, ethical, economic and social uplift of our Society.

We in India need today, more than at any time before, the development and practice of an objective and scientific outlook to replace antiquated, emotional and irrational approaches to our problems; some of these approaches derive from implicit faith in superstition, others are dictates of religion, custom, convention and tradition, and most of them are often in direct conflict with scientific knowledge and an open attitude of mind.

We believe that outlooks in and concepts of life cannot be borrowed in their entirety from the past. They have to be continuously tested, sifted and forged anew by each generation through the practice of ideals, which have their foundation in the ever-growing quanta of knowledge, and through an active tussle with forces opposed to these ideals.

The knowledge gained through the growth of science and technology in modern times has opened up new vistas to mankind. Unimagined possibilities now exist to acquire fresh depths of experience and new ideals, and thereby give a new, richer meaning

* see Chapter II, p. 26 for details.

to human life. This, we believe, is possible only through a large scale nurturing and acquisition of a scientific temper in human society, so that scientific judiciousness and objectivity become universally accepted social and moral virtues.

In this task of "practice what you believe", and adding colour and dimension to human life, we invite you to participate.

Yours sincerely,

P.M. BHARGAVA	**S. DHAWAN**	**A. RAHMAN**
Regional Research Laboratory	Indian Institute of Science	Council of Scientific & Industrial Research
Hyderabad-9	Malleswaram,Bangalore	Rafi Marg, New Delhi-1
India	India	India

For the Foundation Members of the Society

SOCIETY FOR THE PROMOTION OF SCIENTIFIC TEMPER

Object

To bring together in a spirit of tolerance and free enquiry all who are actively interested in promotion of a scientific temper and to take such steps as would further this aim.

Basic premise

Knowledge can be acquired only through human endeavours and not through revelation, and that all problems can and must be faced in terms of man's moral and intellectual resources without invoking supernatural powers.

Membership

The Society is open to any person above the age of eighteen who subscribes to the above basic premise and who pays a subscription of Rs.5/- per annum.

Statement

Science and technology are an integral part of the social development of a country. It is, however, not commonly realized that the dependence of society on science and technology also implies a change in the outlook of people. Science and technology cannot be supported by a society whose members have an anti-scientific outlook.

The technical and industrial transformation, which has given dignity to human life and opportunity for constructive work to the average citizen, also demands a corresponding social outlook and standard of behaviour. This involves the new scientific method of arriving at truth and basing social decisions on these conclusions. There is nothing sacrosanct about this method, but it has a judiciousness characteristic of its own. It is based on observation, collection of data, rigorous analysis of data, arriving at conclusions based on the previous steps, and testing the conclusions in practice to arrive at larger generalizations.

A unique feature of science is that it not only gives a picture of things as they exist but also serves as take-off points for the future. This mode of thought which concretizes human experience, takes into consideration the probability of events and suggests a quantitative correlation between the effort and the achievement.

The countries which today are trying to bring about a technological transformation of society have, therefore, a two-fold objective: changing the outlook of men and changing their environment.

Indian scientists and technologists are playing their part changing their environment but have shirked their responsibility in changing the outlook of people, hoping it will change in the wake of other developments. India is faced today with the problem of breaking down superstition, deep-rooted prejudice and a narrow outlook, which have been keeping people pre-occupied with petty issues. The existing intellectual atmosphere, tainted as it is with these, has led to frustration due to lack of outlet for the creative energies of people.

The situation can only be remedied by giving a scientific perspective to people, so as to organize their effort and channelise their energies towards constructive work. The scientists of this country have no other alternative but to take the challenges and make an effort for changing the outlook of people, to create in people what Jawaharlal Nehru called the scientific temper.

This is also important for the healthy growth of science. Unless there is an informed body of public opinion on scientific matters, the scientists tend to get isolated and form a closed circle. They tend to occupy themselves entirely with their specialized fields and do not tend to take a broader or synoptic view of science. This makes them timid socially, whereas science, by its very nature, is revolutionary.

The popularization of science and the creation of a scientific temper in the country, is, therefore, of the utmost significance and importance for the healthy growth and promotion of science in the country. Once this is effected on a large scale, India will imbibe the spirit of science and be able to make a major contribution to science itself.

India has a long tradition of humanism and has endeavoured to synthesize human values from different traditions and civilizations, and this may contribute something fundamental to the purely technical tradition, which we have taken over from the West. Would this not be a major contribution to world civilization, leading to the creation of new values in science?

Constitution

1. The Society shall be called the Society for the Promotion of Scientific Temper.

2. The objective of the Society shall be promotion of scientific attitude and associated activities in India and for this purpose the Society shall have powers:
   - 2.1 to organize lectures, meetings and conferences;
   - 2.2 to publish proceedings, journals, pamphlets, books and publications;
   - 2.3 to cooperate with other clubs with similar aims;
   - 2.4 to take such other action as may be considered advisable for the promotion of scientific outlook and associated activities.

3.1 The Society shall consist of Honorary, Ordinary and Associate Members and such other classes of members as prescribed in the rules, to be elected in such manner and on such terms and conditions as prescribed in the bye-laws.

3.2 Ordinary members of other societies and other associations which have entered into cooperative arrangements under article 2.3, shall enjoy the rights and privileges of Ordinary Members on such terms and conditions as prescribed in the Rules.

4.1 The Society shall have a President, two or such number of Vice-Presidents as may be prescribed in the bye-laws, a Treasurer, a Secretary and one or more Joint Secretaries as may be prescribed in the bye-laws, who shall be the office bearers of the Society and shall be elected in such manner and shall hold office for one year or for such period as may be prescribed in the bye-laws.

4.2 There shall be a Council consisting of the Office bearers and eight or such other number of members or representatives of cooperating clubs and association, as may be prescribed in the rules, who shall be elected in such manner and shall hold office for such period as prescribed in the bye-laws.

4.3 The Council shall have power to make and amend Rules in accordance with the Constitution and subject to approval by a General Meeting of members of the Society and to frame and amend bye-laws provided that all bye-laws and amendments to bye-laws shall be reported to a General Meeting at the earliest opportunity and shall be subject to the veto of the General Meeting.

5.1 The management of the Society shall vest in the Council which shall have power to take all necessary action for the furtherance of the object of the Society.

5.2 The Council shall meet on such occasion as may be considered necessary by the President, and also on the written requisition of not less than three members of the Council.

5.3 The Council shall be subject to and bound by the directions of the ordinary members of the Society assembled in a General Meeting convened in such manner as prescribed in the bye-laws.

6.1 A General Meeting of the members of the Society shall be convened at least once every three years to consider the report and audited accounts, to elect the required number of office-bearers and members of the Council and to transact such other business as may be necessary.

6.2 Special Meetings of the members of the Society may be convened by the Council when necessary and shall be convened on the written requisition of not less than one-fifth of the members on the rolls of the Society.

6.3 The General Meeting of the members of the Society shall be the supreme controlling authority and shall have power to amend the constitution by a two-thirds majority of members present at the General Meeting subject to confirmation by a majority of members present at another General Meeting to be held not earlier than one month after the General Meeting at which the amendments were adopted.

Provisional Rules

It was resolved to adopt the following Provisional Rules for two years or until modified by a General Meeting of the members of the Society.

7.1 The membership fee of Ordinary Members shall be Rs.5/- per year.

7.2 A member paying a consolidated fee of Rs.150/- shall be entitled to become a Life Member of the Society.

7.3 Members of the cooperating clubs, and associations shall enjoy the rights and privileges of an ordinary member of the Society on payment of a fee of Rs.2/- per year.

APPLICATION FOR MEMBERSHIP
SOCIETY FOR THE PROMOTION OF SCIENTIFIC TEMPER

I am interested in the promotion of scientific outlook in India. I would request you to enroll me as a member of the Society for the Promotion of Scientific Temper. I enclose Rs.5/- (Rupees five) being my membership contribution for the year.

Name ....

Educational status ....

Occupation/Profession ....

Address ....

I SUBSCRIBE TO THE BASIC PREMISE OF THE SOCIETY:

KNOWLEDGE CAN BE ACQUIRED ONLY THROUGH HUMAN ENDEAVOURS AND NOT THROUGH REVELATION AND THAT ALL PROBLEMS CAN AND MUST BE FACED IN TERMS OF MAN'S MORAL AND INTELLECTUAL RESOURCES WITHOUT INVOKING SUPERNATURAL POWERS.

Signature

To
The Society for the Promotion of Scientific Temper
Post Box No. 237, New Delhi–1

V

THE SCIENTIFIC OUTLOOK*

P M Bhargava, K T Achaya and *Bharat Bhushan*

While modern science and technology is now accepted everywhere as an integral part of one's everyday living, few ever stop to think, particularly in our country, about the obligations which rest on the users of the benefits of science and technology. The most important of these obligations is the understanding of the scientific method and the development of a scientific outlook. In this article an attempt will be made to detail the need for such an outlook, to define some of the problems in its creation and to make some suggestions in regard to the responsibilities of the scientists in the development of a scientific outlook.

What are Scientific Method and Scientific Outlook?

The benefits of science with which humanity at large are familiar, are seldom recognised by the layman to be a result of a simple, systematic, well-defined and objective approach: the application of the scientific method to the solution of problems and discovering truth. The scientific method stands in direct contradiction to the way of religion, dogma and faith which is based on the premise that truth can be revealed and which was the only approach available for

---

* This article appeared in *Campastimes* (IIT, Madras), February 1969, pp.47-49.

solving problems and discovering truth till a few centuries back. The scientific method rejects revelation as a means of discovering truth and substitutes it by the technique of observation followed by careful experimentation and logical deduction. Thus, in contrast to revelation which is highly personal, the scientific method is universal.

It is only logical that once science has permeated every level of human existence, the method of science must become a way of life and science must be conceived not as knowledge of facts but as a way of thinking. An unqualified acceptance of this role of science, defines 'scientific outlook'; it is a corollary of acceptance of the scientific method as the only way of discovering truth.

Need for a Scientific Outlook

It may be perhaps argued that a scientific outlook is necessary only for those who practice science and that such an outlook is of little use to a person who is normally concerned only with the utilization of the fruits of science and technology and not with science and technology per se. There is no greater fallacy than this. Today scientific answers are either available, or there are reasons to believe (from the trend of modern researches in chemistry, biology and physics) that such answers will be found in the future, to virtually all the major questions which humanity has been asking itself since man came to be endowed with intelligence. In fact, by providing answers (or opening avenues for the acquisition of such answers) to common questions such as those pertaining to atmospheric or cosmological phenomenon (e.g., rain, thunder, day and night, and eclipses), or to the more sophisticated questions such as the nature of the universe, the origin of life, the mechanism of heredity, the cause of disease, and the basis of the various physiological processes (including those which control behaviour and brain function), science has established the validity of its method and has thus given a new technique to humanity for solving its problems. Attempts to provide answers to these questions in the past were based on the teachings of religion, dogma, faith, custom, convention and tradition. Science has ultimately proved to be right in every case where its results and conclusions have differed from the teachings of religion, etc. Darwin's theory of evolution propounded a little over a hundred years ago was vehemently contested by the Church as it went against its teachings but is today fully accepted even by the Church.

In view of such astounding successes of science, particularly where its results come in conflict with religion, dogma or the like, it can be assumed with confidence that a scientific outlook, which is based on an understanding of the method of science, can help arrive at a rational solution, in terms of human resources and knowledge (and without invoking any supernatural power) of various problems which face man as an individual or as a part of the social community. In fact, a scientific outlook is today a prerequisite for proper appreciation and pursuance of the ideals of liberty, freedom and self-reliance and, in a modern setting, it can be a strong pro-secular force.

Even if one should like to be concerned only with the benefits that science brings to humanity, one would immediately see that a scientific outlook could lead to a better appreciation of the impact of science and technology on society and thereby provide the right kind of atmosphere in which science and technology can flourish and their importance recognized. Moreover, such an outlook is essential to ensure application of science and technology exclusively for the benefit (and none for the destruction) of mankind. The presence of a scientific outlook amongst the masses is necessary for developing in the people the desire to understand natural phenomenon and thus to encourage acquisition of fundamentally new knowledge on a large scale. There could be no two opinions about the fact that the whole history of the progress of mankind is a history of the acquisition of such knowledge.

Problems in the Creation of a Scientific Outlook

Perhaps the greatest stumbling block in developing a scientific climate is the widespread belief in superstition and in the supernatural as a result of uncritical acceptance of the dictates of religion, dogma, faith, customs, convention and tradition. Scientists and intellectuals are no exception to this, particularly in our country. This dichotomy amongst them is of much greater concern than the belief in superstition or supernatural on the part of a person who has had no access to the fund of modern scientific knowledge. Illiteracy and lack of proper education amongst the masses accentuate the problem. Emphasis should be laid on the word 'proper'; a large number of existing textbooks in this country, instead of propagating rational, scientific thinking, do just the opposite. For example, the bulk of our school and college students cannot discriminate between what is legend and what is history. It is not uncommon to find a highly educated person, even a Ph.D. in science, telling you the exact date of the rule of Rama in the country! Lastly, one faces the lack of a policy and of adequate media for dissemination of scientific information.

Steps Which Should be Taken for the Creation of a Scientific Outlook and the Responsibility of Scientific Workers in this Respect

It is of the utmost importance to make the teaching of science compulsory through schools, and to stress its principles and methods while teaching it. Science teaching must also emphasise the role of science in everyday life, for example, in an Indian village. This would involve a complete revision of the syllabus (which should take into consideration the recent developments of science), the development of proper textbooks with adequate provision for frequent revisions, a programme of training of teachers to teach science, and creation of adequate facilities in schools for such teaching. It is clear that this cannot be achieved except through the State and Central departments of education, but the direction for this could and should be provided by a suitably organised body of scientists.

Efforts to improve science teaching should be supplemented by efforts to popularise science. Competent scientists should devote time to write articles in newspapers and help set up a system which would ensure that science reporting in newspapers is true and interesting instead of sensational and prosaic as it mostly is at present. As many popular science journals as possible should be started; for example, there can be one in every institution of the level of a high school or above.

Scientists in positions of influence could see that greater time is devoted on the radio and television to popularise science. At the moment, these two media are doing virtually nothing in this respect.

State Governments should be persuaded to establish at least one science museum and one museum of natural history in every State. If such museums are established, all school and college students must visit these museums at periodic intervals, and facilities for such visits, which should be made compulsory, should be provided by their respective institutions. The museums should stress the principles of science, the joy of scientific discovery, the substitution of irrational with rational thought, and the importance of deductive thinking.

Another way in which scientists could help in the creation of scientific outlook is through the establishment of science associations formed with the express objectives of (a) spreading consciousness of the aim, method and values of science; (b) removing belief in the supernatural and in superstitions; and (c) removing dichotomy amongst scientists themselves.

Lastly, we think the time has come when every scientific worker—and we include students of science and technology in this category—must do a little bit of introspective thinking with the objective of emancipating himself from all that is anti-science. The basic methodology one learns when one studies science and technology can lead him to only one logical conclusion: the decision to renounce faith in revelation and to apply the scientific method to his everyday living. There cannot be any substitute to the exemplary behaviour of scientists themselves as an aid to the propagation of a scientific outlook.

VI

INDIAN SOCIETY AND THE SCIENTIFIC TEMPER*

P M Bhargava and *K T Achaya*

What is Scientific Temper?

Scientific temper was a phrase much in Jawaharlal Nehru's vernacular. He reiterated it not only in speaking of science, but also in exhorting his countrymen in diverse contexts. The phrase is an attractive one and has both brevity and comprehensiveness, for *temper* indicates all the hues of man's thinking, nicely qualified to the plausible and rational with the adjective *scientific*. We might pause to define the phrase, or at any rate to envisage its implications for the citizen. It would imply certainly a willingness to consider *all* facts and not merely facts which are in consonance with one's own thinking or comfort. Going further, it would mean an active search for such information by study and questioning. It would also imply a trust that events are shaped by the fruits of man's labour, and a healthy skepticism towards all claims of supernatural participation in his affairs. In fact, the scientific attitude is simply one of an adherence to facts, an ability to revise opinions and a rational skepticism to claims for non-material intervention.

---

* This article appeared in *Science Perspectives*, A Rahman and K T Achaya (Eds.); Academic Books, Bombay, 1969, pp. 19-25.

Nehru's Vision

Development of a scientific attitude among the people was an important part of Nehru's vision of India. He recognized, however, the extent of the transformation required of contemporary Indian society before his vision could materialize, and this was sufficient to despair even an optimist like him, for a scientific temper is conspicuously lacking in the country, even among those with an ostensibly scientific training. Scientists lay aside the mantle of incredulity and deductive logic when they get home and kick their shoes off, relaxing into every kind of obscurantist fad and fallacy. Doctors still see no contradiction in their patients visiting Tirupati or the local temple for cure of physical ailments; indeed they do so themselves. Modern agricultural scientists have little conviction of the benefit of inorganic fertilizer in their kitchen gardens; fertilizer is something to be doled out to the farmer, or to be used in speeches.

Daily newspapers and magazines yield rich dividends to the seeker after obscurantism. Practically every major newspaper has an astrological corner with predictions for the week to follow: "You could come into a large sum of money this week. The health of your family members may require your attention. You may meet influential people who will help you": modern India's oracles of Delphi! The child of a scientist will still get married in the small hours of the morning so as not to offend the planets in their whizzing courses. Practically every scientist will take a day or two off in the year to perform the annual ceremonies of his dead parents and, if questioned will blame the insistence of some relation who is well out of reach. There are auspicious days for travel and certain specially lucky days and numbers. Where in the midst of this welter of irrationality, even among scientists, is the hope for the creation in the people of a scientific temper, an open attitude of mind rooted in the questioning of dogma and authority? Is the attempt possible, or even worthwhile?

No Other Way Out

The answers have to be in the affirmative. Indeed, the questions are simply rhetorical; there is just no other way, whether for the scientist or any other Indian, for the scientific attitude as a way of thought of people—scientists and non-scientists alike—is not merely desirable but essential if we wish to have a strong base for secularism in the country (since antisecularism or parochialism is the very negation of the scientific attitude), if we wish to find quick, permanent solutions to the multitude of problems we are faced with such as food, public health and family planning, and if we wish to raise the living standards of our people. How, then, can we effect a large-scale promotion of scientific temper? And what have we done in the past and what are we doing now in this direction?

What Are We Doing?

India has not had a tradition of objective, rational and scientific thinking. Nirad Chaudhuri has pointed out in an article in *Encounter* a year ago that even the coldly-sober *Arthasastra* of Kautilya seriously lists in the last chapter the methods and drugs which spies should use to become invisible, and the otherwise scholarly *Laws of Manu* describes the appropriate sorts of rebirth for each worldly sin. India never went through the renaissance, which engendered the rationalist tradition in Europe. The industrial revolution that had established itself in Europe for decades was just beginning to creep into this country when the Independence movement started. The latter, however, soon became strongly linked to revivalism through the personal beliefs of Gandhi. There can be little doubt that by this means he was able to mobilize enormous numbers to the national purpose, but he also set in motion various factionalisms, which remained submerged during the intense drive for Independence, but have reared their ugly heads since then. Nehru represented a positive gain for independent thinking, setting himself up against all types of reaction. One remembers with gratitude the strong ridicule he heaped over the absurd *Ashtagraha* episode. He was the only national leader to do so, but we suspect that the impact his ideas had on the country as a whole was probably small. While it may have set at bay some obscurantist and religious forces, these were only biding their time, and their resurgence was markedly noticeable even in the brief regime of Shastri. The personality of the present Prime Minister, while certainly of rationalist conviction, has not yet had time to make its impact.

While such father figures undoubtedly influence the outward manifestations of rational thinking in India, a greater permanent effect must derive from the impact of modern technology. It is impossible for a worker in a factory to insist that his neighbour should not be an untouchable. Large-scale industrialization, already begun, should eventually prove to be an important aid in the development of a scientific temper among the people of the country. There have been other efforts too. Organized bodies have come into existence with rationalist aims, and without political overtones. These may at present be drops in the ocean but deserve passing comment. The Indian Humanist Union was founded in 1960 at Lucknow and has just started a small journal. Its purpose is to propagate rational and scientific humanism within a tolerant religious affiliation. The Society for the Promotion of Scientific Temper arose out of discussions in Hyderabad in 1964 by a group of scientists drawn from many nations, which were fully reported in an issue of the intellectual monthly, *Seminar*. There are branches in several cities in the country trying to spread rational belief, and take a stand against the fetters of traditional religious dogma. There are other indications of ferment in the country. Student unrest, and the frequent disturbances of law and order, are not merely signs of indiscipline but embrace an element of genuine questioning of established norms and a groping towards values which are preached by their elders but never practiced. There are reasons, therefore, for viewing the future with optimism; the first steps have been taken, and one can only hint that the changeover will not be cataclysmic.

Responsibilities of the Scientist

We may now consider briefly the responsibilities of the scientist and the educated non-scientist in respect of the promotion of scientific temper in the country. Both are of course part of the common social fabric of Indian society, and if a distinction is made at all, it is only because the scientist uses in his work, or at any rate is expected to use, a certain methodology as a matter of course. On the other hand, while the layman doubtless conducts his everyday life on some rational basis of cause and effect, the degree to which he does so is largely a matter of individual predilection rather than of training.

We have stated that scientists themselves often display a lamentable lack of a scientific attitude of mind as soon as they leave the laboratory. The business of day-to-day living is considered a thing apart, subject to different laws and rationalities. We have earlier cited some examples of this. A scientist may be expected to use the analytical and deductive method of reasoning to a greater degree than his lay neighbour in the small matters of everyday living, but with the larger issues of society and its values he is generally reluctant to concern himself as not being his business. There is some point in this. The parameters involved in such wider questions are so many, so varied and sometimes so intangible that the attempt may be abandoned even before starting. A scientist may doubt whether these factors can even be enumerated, and assuming that he was able to do so, whether any kind of opinion will at all be possible. And yet a scientist owes it, at least to his neighbours and friends, to make the effort, for it is to him that they will look for an opinion on questions having a direct or indirect bearing on science. His conduct and behaviour will, therefore, unconsciously set a pattern to his circle of friends. In this respect a scientist occupies a privileged position; he should realize it and accept the responsibility implied in it. Most Indian scientists should cultivate a much greater constructive interest than they have so far taken in the policies of the government in such matters as science teaching, technical education, outlays on research and the like. This is obviously a difficult task. Being a good scientist just in the professional sense is in itself a full-time job. There is not only the constant grind to keep up with the avalanche of scientific literature in one's own field, but also the distasteful but necessary business of conveying what one has found in one's work to one's scientific peers, without which the pursuit itself will be of little avail: all the awful business of writing research papers, preparing reports for all and sundry, sitting through seminars and conferences and the rest. And yet the scientist must accept his heightened social responsibility, if only for the reason that the knowledge he has helped to acquire may be misused to his own great detriment.

There are important topical questions such as the making of the atom bomb in India, or the outlay on science and technology in the next plan. These are matters on which he should have an informed opinion based on analysis and deduction of factual knowledge. He must take the trouble to obtain and study this evidence, and to reason out and derive his attitudes, whatever they may be. The scientist should not forget that science in the larger sense is today the only activity, which can create true wealth. It has come to be the prime mover of society and world affairs, changing radically our basic attitudes and our concepts and values.

For the educated but professionally non-scientist Indian population, vast in numbers if not in percentage proportion, the challenge lies in developing the willingness to work out one's basic concepts for oneself, discovering and eschewing those which are in conflict with the demands of obvious commonsense and reason. This would be the primary debt that they owe to education which, after all, should be training in rational thinking and in inculcating a scientific outlook. There is no difficulty in this; all that is required is a conscious effort. It is true that the day-to-day worries and problems of living, rising prices, concern for the health of one's family, and the education of children, are all serious, stultifying forces in the true exercise of the mind and reason, but the effort will simply have to be made if life is to have any meaning at all. A man's opinions are derived from many sources. His basic ethical values are probably those derived from early religious training. There may also be the beliefs and values cherished in his own class, group or family. Experimental studies have shown that the attitudes and concepts, which he develops later, will reflect his discussions with friends, and all-too-often will be simply the opinions of someone of strong personality whom he admires. Anyone who thinks, reasons and attempts to find out facts will dominate a group of non-thinking people. One of the ways, therefore, in which a man can develop his own rationalities, is by conscious study and debate with himself. He would also benefit by energizing and activating any discussion group in the society to which he belongs and thus airing his opinions for discussion. It is better to have an opinion, however wrong, than no opinion at all. The important thing is to give up passive attitudes, and keep one's mind active.

Only in this way can we expect to develop an informed society, which is our only long-range guarantee of a better and more rational life for our people.

The scientific or rationalist attitude comes with maturity, but a beginning must be made early. A practical question is how an average citizen, when of rationalist conviction, can inculcate ethical values in a child without bringing in the easily understood concept of an all-powerful God. Religion is woven into the warp and weft of every living culture; a child is certain to be exposed to it in its daily living. We believe that rationalism and reason rather than fear and placation of the unknown must and can be made intelligible to a child. This admittedly requires more patience than leaving the job to God, but the parents will discover that it is in the final event most satisfying, both to them and to the child.

Conclusion

The exercise of scientific temper or scientific humanism is not relevant only to the denizens of a sophisticated and intellectual nation, to whom the fruits and perils of science are obvious. It has perhaps an even more important role to play in an emerging nation. It can induce thought, awaken society from the apathy into which it has fallen, and make the vast population a moulding and creative force instead of a crushing dead weight. In this transformation, every educated Indian has a special part to play. This is the hope—our only hope—for the future.

VII

OBSCURANTISM AND ACADEMICS*

P M Bhargava

According to dictionary, 'obscurantism' means opposition to enquiry or reform. In common parlance, it has come to mean opposition to science or to truths discovered through science (as spirit of enquiry is the main motivation behind science), and to social and economic progress (as continuous reforms are the basis of such progress). Today, obscurantism may be considered to encompass all acts of irrationality such as communalism, unscientific ideas and beliefs, superstitions and dogmas, and blind adherence to religion, customs, convention and tradition.

Origin of Obscurantism

Beginning of Religion

Researches in biology over the last one hundred years or so, beginning with the enunciation of the theory of evolution by Charles Darwin, have shown that life must have originated on this earth from non-living materials. The earliest, primitive forms of life were probably small and simple, and capable of random change which, once effected, was inheritable. Out of the changed forms, or 'mutants', in

---

* This article appeared in *The Secularist*, July-September, 1970, pp. 7-13.

the progeny, those which were able to adapt themselves better to the environment had greater survival value. With this advantage, in many cases, they made the parent forms extinct in the course of time. In this way evolved the multitude of living forms, present and past, like dinosaurs and bacteria, fish and man.

During evolution, at some stage, man came to be endowed with intelligence. Since intelligence is in a way synonymous with the ability to ask questions and to find their answers, man must have asked himself a progressively increasing number of questions when, at the dawn of human history, he came to possess this gift. It would be safe to surmise that these questions would have initially pertained to what he could see around him: (a) the non-living objects, (b) the living objects, and phenomena like birth, reproduction, death, etc., (c) physical phenomena such as heat, light, lightning and sound, and (d) extra-terrestrial phenomena and objects. It must have been an intrinsic desire in him to find answers to questions such as these which made him gullible in accepting 'revealed' truths. This combination of curiosity and credulousness probably contributed, in a large measure, to the development of myth, legend, superstition, belief in the supernatural, and the concept of God or gods. Organised religions (pagan or other) and systems of philosophy based on them must have subsequently evolved—and been accepted—largely as a result of the codification of these developments. The fact that the major religions of today differ primarily in their dogma and not so much in their moral or ethical content, would support the above argument.

Till about five hundred years ago, the teachings of religion were rarely, if ever, questioned. Religion was, till then, believed to be the only way to acquire new knowledge; its method was revelation, and this implied that truth could only be found through divine intervention. History tells us that any attempt to deviate from the dictates of religion or reluctance to accept them has caused untold misery to those who dared to do so.

Method of Science

About the fifteenth century after Christ, the doors of another way for acquiring knowledge were opened: the way of science which is based on observation and experimentation, and which calls for an unbiased analysis of the observed and the experimentally derived facts, an unequivocal rejection of what is not verifiable, and readiness to modify or reject a theory if newly discovered facts did not fit it. The Renaissance in Europe, followed by the scientific and industrial revolution, was largely a consequence of the recognition of the method of science as a technique which could provide answers to questions and thus lead to truth.

In the first stage of the scientific revolution, the protagonists of the new method focused their attention on questions pertaining to extra-terrestrial objects and, phenomena: we had Galileo, Copernicus and Newton who toppled theories about the earth and the sun supported by theologians for centuries. The emphasis then shifted to understanding the nature of the non-living world around us. In the last twenty-five years, the emphasis has been on biology, that is, on questions pertaining to life.

The outcome of all these efforts has been phenomenal. Today, many major questions about non-living objects, about phenomenon concerning life, and about extra-terrestrial objects and phenomena, which have intrigued mankind for millennia, have either been answered, or the way paved for answering them. The five elements of ancient mythology—earth, air, fire, water and ether—have given way to the ninety-two natural (and several artificially made) elements, of which—we now know—the entire universe is made. We generally understand the basic nature of heat, light, electricity and magnetism. We no longer wonder at the variety in the materials around us, as we understand how they are derived from the ninety-two elements by chemical combination between them: we largely understand the nature of this combination. Most of us do not any more worship the sun, the moon and the rain god; and the movement of planets in the sky, the eclipse, or the periodical appearance of a comet, are no longer mysteries with supernatural connotation.

We now also understand the nature of materials—their chemistry—of which all living forms, including man, are made. We know a great deal about what makes us tick—the eye-lids blink or the heart beat. Our knowledge of nutrition and our ability to conquer disease has led to a considerable increase in the life-span of man. We today know why the progeny of man is man, and that of monkey, monkey. We understand what it is that makes us what we are, why two identical twins are so similar, why two brothers are a little less similar, why even unrelated human beings have so many things in common (although they are less similar to each other than two brothers), why there is some similarity between monkey and man, and why the fish are so different from man or amoeba. We have solved the mystery of heredity and of the origin of the immense variety of living species around us. We have also gained considerable understanding of death. We can today make a man "die" by taking his heart out and then make him come alive again by putting someone else's heart in him. We can take parts of him, organs or cells, after he is dead and then keep them alive, perhaps for ever; we can make our body cells multiply in the test tube, and our great-great-grandchildren can see them, pulsating with life, a century after we are dead. Modern biology has made the border line between life and death very tenuous on a strictly materialistic basis; the concept of soul has become redundant and the theory of resurrection *or* reincarnation untenable.

There are of course questions we have not yet been able to answer; for example, we do not yet know the cause of cancer, and we do not understand the mechanism of storage of information in our brain, or of the recall of this information (that is, memory). We can nevertheless predict that answers to questions such as these will be found by science, perhaps in the not-too-distant future.

Obscurantism Appears

Before science appeared on the scene and came of age, religion (and related activity) had already laid down some answers to questions such as those mentioned above. Obscurantism started to manifest itself when people began to challenge these answers which were largely based on revealed truth. As it turned out, the revealed truths and the answers given by science were at variance with each other: very often, completely irreconcilable. For at least four reasons, the answers obtained by the application of the method of science were considered more acceptable when they come in conflict with the answers given by religion. First, incontrovertible evidence was available for the scientific truths; secondly, they could be verified; thirdly, one could make accurate predictions based on them; and lastly, no matter who asked the question the same answer was obtained. This was not so in the case of answers provided by religion and philosophy. The evidence had usually not been incontrovertible and the proofs never rigorous. The conclusions were not experimentally verifiable. The answers had varied with the religion, or the system of philosophy or even the individual's interpretation of them. And neither religion nor philosophy could make predictions which came true. Therefore, whenever science came in conflict with religion or the like, it was science which won, and religion had to modify its theory or dogma; there is not a single case in history to the contrary. As the knowledge acquired through science grew, the fallacy of the revealed truths—the answers provided by religion—became increasingly clear, and the superiority of the method of science came to be progressively established.

Unfortunately, however, not everyone accepted what was shown to be true by science, because scientific knowledge has not been accessible to the vast majority of human population. Thus, while science progressed by leaps and bounds through the efforts of a few, most of the others who were not a direct party to this progress, did not accept the validity of scientific discoveries. It was this division which put obscurantism into focus. If there were no science, there would have been no obscurantism. With the development of science, obscurantism too achieved greater definition, and today, when science and technology constitute the fabric of modern society and are the most important tools of progress, the forces of obscurantism stand out, by contrast, as the most important anti-progressive factor. The majority of mankind still clings to the "older knowledge", at least partly because the arguments of a scientific discovery are not comprehensible to the average person. Obscurantism has been thus a logical outcome of progress in a society, a majority of whose members did not understand the basis of the progress. In other words, today's obscurantism is a consequence of a situation where only a small part of the society is "knowledgeable" in the true sense of the word (I am here restricting my definition of knowledge to what is acquired by the application of the scientific method).

Responsibility of Academia

Knowledge and Obscurantism

There is a well-known saying by Tennyson: "More things

are wrought by prayer, than this world dreams of". Perhaps the more appropriate saying in the modern context would be: "More things are wrought by knowledge and action, than this world dreams of". Problems may be solved by knowledge, not by prayer. Prayer is an obscurantist practice and, as we have seen, obscurantism and knowledge do not go together. No obscurantist practice can cure typhoid or malaria; it is only our knowledge of the origin of these diseases that we have acquired through science, and their cures similarly discovered, which allow us to combat them successfully. The successful eradication of many diseases in several parts of the world has not been achieved by prayer but by human effort based on knowledge acquired through science. Similarly, in areas, where droughts prevail; or no vegetation is possible on account of lack of water, one is able to get good crops year after year, not by soothsaying but by deliberately planned irrigation schemes, based on the knowledge of certain facts. Against the background of the sum-total of such knowledge and experience, gathered by human effort from the time man came to be endowed with intelligence, obscurantism has today no place in human society; it becomes at once redundant and harmful.

Education and Obscurantism

The link between knowledge and obscurantism makes it clear that education must play an important role in ridding society of obscurantist ideas, as one of the primary purposes of education is to impart knowledge. It follows that in a society like ours, which has been primarily obscurantist (and in which the pressures of customs, convention, tradition and religion are such that if one is left to develop from childhood to adulthood without any counteracting influence of knowledge, one is likely to develop obscurantist views), the first ones to be emancipated from the shackles of obscurantism should be the educated. (Indeed, one of the ways, in which the educated in our country can discharge their obligation to society is by making use of the knowledge they have acquired to shed obscurantist ideas.) Even amongst the educated, one would expect that the first to shed obscurantism would be those whose profession is education, that is, the academicians. While one could, in a society like ours, understand (though not condone) the adherence to obscurantist views by the multitude of the educated, there cannot be any justification whatsoever for the adherence to obscurantism by our academicians. Further, since the logical basis of the refutation of obscurantist ideas is largely scientific, one would justifiably expect that from amongst the body of academicians, the first ones to realise the invalidity of such ideas would be the scientists. They are the ones who are directly and professionally concerned with the physical nature of the universe and with questions pertaining to life—areas which have given rise to the largest proportion of obscurantist ideas.

The Situation in India

In India, unfortunately, the situation in respect of the attitude to obscurantism of either the educated, or amongst the educated, of the academicians, or amongst the academicians, of the scientists, is very unsatisfactory. A vast majority of scientists are prone to hold obscurantist views. They believe more or less blindly in the dogmas of religion, in the teachings of classical philosophy, and in custom, convention, tradition, etc. They accept and propagate all kinds of fads, for example food fads; they are unable to distinguish between legend and history (many university teachers and research workers would be unwilling to accept that Rama, Lakshman, Sita and Krishna, or various incarnations of Vishnu, are legendary figures); they have implicit faith in miracles (they, would consider you a heretic—an undesirable person to associate with—if you said that Jesus Christ or the Buddha could not perform miracles any more than you or I can). It is not knowledge (which it is their business to acquire) but superstition and irrational and illogical beliefs which guide their lives. Most of them still look for an auspicious day to travel. They regularly practise rituals like visiting a temple, mosque or church; the Hindus believe propitiation of one of their numerous Gods, for example, Lord Venkateshwara at Tirupati, would materially benefit them, or that having a dip in the waters of river Ganges will wash off their sins. Much of their valuable time is spent in everyday recitation of prayers or in worship at home; this is often done even at school. They have a blind belief—without any scrutiny—in the supernatural powers of individuals like Sai Baba. They believe in astrology and in irrational medical practices like homeopathy, which go against the very grain of science. They support unscientific and irrational movements like the one for a ban on cow-slaughter. They are reluctant to accept, at least in private life, well-established scientific theories like that of the origin of man as a result of evolution by natural selection; they would still like to believe that man was deliberately put on this earth by a Supreme Being. They believe in the existence of soul and in life after death. They are wedded to the concept of an all-powerful deity like God and consider it a sacrilege to challenge it. They would rather spend their limited resources on the celebration of religious festivals than on the education of their daughters, and in the heart of their hearts they nurture sectarian views. It would indeed be a rare scientist in our country, whether in a school, college or university, or in a research laboratory, industry, or the Government, who would not be prone to obscurantist views such as the above; by and large, he would reject conclusions based on the findings and discoveries of science and show preference for revealed truth.

As a consequence of all this, our academicians—

scientists included—mostly look backwards. for inspiration, and not forward. One can, therefore, conclude that most of those who technically come under the definition of academicians or scientists in our country, are in reality, neither. This helps to perpetuate obscurantism as such views on the part of those engaged in an academic profession, only help to consolidate obscurantist ideas amongst the masses.

The Solution

Proper Education

The only permanent, long term insurance for getting rid of obscurantism from our society is to lay emphasis on the *right kind* of education. One of the professed objectives of education must be to equip the recipient to fight obscurantism of which he would otherwise be a victim. Education, right from the very beginning, should be science and knowledge-based. Today, textbooks are one of the prime conveyors of obscurantist ideas to school children. They should be carefully scrutinised before they are prescribed by an independent group of "committed non-obscurantists", and such people should be encouraged to write textbooks themselves. Further, every possible attempt should be made to emancipate teachers from obscurantist ideas, for example, by holding for them seminars, summer schools, refresher courses, etc., in which the emphasis is on anti-obscurantism. The general approach to education should be a positive one, that is, to create an environment where a student or a teacher would be able to see for himself that obscurantist ideas are not only incompatible with modern knowledge but that they generally act as a great obstacle to his own personal progress as well as that of his fellow human beings.

Role of the Government

The Government should take strong exception to

obscurantist teaching in schools, for example, by stripping them of their grants and recognition. The Government should encourage its own employees to take positive steps towards fighting obscurantism. One might argue that this would be against the professed pro-secular outlook of the State. Such an argument would, however, have little validity. Even today, for example, if a Government employee refuses to travel in a certain direction on a Wednesday because it is against his belief (he can cite religious books, custom, convention and tradition in support of his belief) he is liable to disciplinary action by his superiors. The time has come when we must realise that if any action based on the dictates of religion, classical philosophy, dogma, custom, convention or tradition goes against the basic, long-term legitimate interests of the society at large, those in power must take steps to ban such action. Our own history supports this argument. For example, the Sarda Act and the Act permitting the remarriage of Hindu widows were, at the time they were passed, considered an infringement on the rights, customs and religion of the Hindus; today they are accepted in a large measure by the Hindus themselves, A stern anti-obscurantist policy by the Government would, therefore, not be anti-secular inasmuch as it would not imply discrimination on the basis of religious or allied considerations. I can conceive of many legislations which if passed would help the anti-obscurantist movement in the country to a great extent, and thus make the task of national progress immensely easier. Unfortunately today even high dignitaries of the Government support and encourage in official speeches—to cite one example—an irrational and unscientific medical system like homeopathy. A senior scientist who was till recently Scientific Adviser to a Central Ministry and occupied one of the top six Governmental scientific positions in the country (thus presiding over the destinies of thousands of scientific workers) had, on the invitation card for the wedding of his son, the photograph of Satya Sai Baba (who is, at best a good conjurer) in whose "divine" presence the wedding was to be solemnised; and a Minister of the Government of India shied away, at the last moment, from presiding over a serious-minded academic conference on obscurantism.

Press and Publicity Agencies

The press and Governmental publicity agencies like radio, television and the State departments of information, can be highly effective in fighting obscurantism in the country. Unfortunately, so far they have been actually serving, directly or indirectly, the cause of obscurantism. One could hardly find an issue of a widely read newspaper or periodical published in the country today which somewhere in its column does not propagate an obscurantist idea.

Voluntary Organisations

Lastly, as many voluntary organisations as possible should be formed and supported by people who are so fortunate as to be emancipated from the shackles of obscurantism themselves. The approach should be positive, that is, propagation of knowledge and the creation of an environment, which would automatically lead people to reject obscurantist ideas by virtue of their own logic. The fight against obscurantism should not be carried out in an evangelistic tone. Indeed, their approach should distinguish these organisations from the numerous sectarian ones in the country, which are causing untold harm to national (and human) interest.

The Situation Elsewhere

The problem of obscurantism is not peculiar to this country

alone. It exists all over the world to a lesser or a greater extent. However, a great redeeming feature of the situation in the more advanced and prosperous countries, for example of the West, is that the intelligentsia, inclusive of the academicians, is by and large emancipated from the shackles of obscurantism. Further, even though the masses are still prone to many obscurantist ideas arising out of religion and philosophy, science and technology have permeated their lives so much that religious dogma is being questioned more and more by the average layman. Thus two of the most widely circulated magazines in the world today, *Time* and *Life,* have in the last few years questioned the existence of God and of soul, supporting the arguments of scientists against unqualified acceptance of their existence, and laying the onus of proof of their existence on the theologians.

Conclusion

Lastly, let us recognise the fact that getting rid of such deeply-rooted obscurantist views as prevail among the vast majority of our people is not a trivial task and is bound to take time. One must not expect quick results. What is important is to *begin* an organised effort which will sustain. The responsibility of making this beginning must lie with the academicians, the intelligentsia and the educated elite of the country.

VIII

THE SCIENTIFIC TEMPER*

P M Bhargava

Some miles away from Bangalore, in a place called Puttaparthi lives a man called Satya Sai Baba who is considered by himself and his disciples as God incarnate. He has acquired a vast following, largely as a result of the miracles that he claims to be able to perform. For example:

(a) Shri Satya Sai Baba can apparently produce material objects, such as Swiss watches, out of nothing and nowhere. If no trick is involved in such an act, it would mean that he possesses the power to convert "non-matter" into matter (a conversion which is virtually impossible on the basis of science under the conditions it is reported to be achieved).

(b) Shri Satya Sai Baba often gives to his followers *bhasam* (or *vibhuti*) which has been blessed by him to have curative and other extraordinary properties. The blessing, it is claimed, results in the amount of the *bhasam* remaining at the original level, in spite of small amounts being taken out from time to time by the devotee. In other words, the *bhasam* cannot only regenerate itself, but also knows when to stop its own regeneration (a virtual impossibility from the point of view of science).

---

* This article appeared in *Secular Democracy*, Nehru Number, 1976, pp. 109-115.

(c) Photographs of Shri Satya Sai Baba blessed by him, it is claimed by his devotees, keep on shedding intermittently, without intervention by any human agency, *bhasam* or *vibhuti* of the same type and properties as those mentioned above. Shri Satya Sai Baba, therefore, wishes to project to his followers that he has the capacity of not only creating matter out of nothing, but also of directing such creation *elsewhere*, that is, at a place where he is not physically present (such a claim contradicts the fundamentals on which science is based).

Millions of people in the country implicitly believe that Shri Satya Sai Baba is not resorting to plain and simple tricks when he performs these miracles. But there is one man, if he were alive, who would have never believed in the so-called godly and supernatural powers of Shri Satya Sai Baba: this man was Jawaharlal Nehru. It was he who put in our vocabulary the now widely used term, the *scientific temper*. Nehru has been one of the few in our country—scientists included—who have held a position of power and influence, and who have realized that the lack of scientific temper amongst the people of our country has been one of the most important stumbling blocks to social progress, to the emancipation of the masses and, indirectly, to economic progress. Nehru was amongst the first in the country to realize that as long as scientific temper was not woven into the fabric of our thinking, the masses may not be able to claim and acquire their legitimate right. He genuinely believed that exploitation of one kind or another by one group or the other would continue in the country and prevent the emergence of a truly socialist, secular and democratic state, unless we developed scientific temper on a mass scale. It was this belief which led him to readily accept the first presidentship of the Association of Scientific Workers of India (ASWI) when it was formed in the late

1940s. One of the major objectives of ASWI has been the promotion of scientific temper. Unfortunately, this organization has not fulfilled the promise with which it was started, and the blame for this failure lies squarely on the shoulders of our scientists.

II

Scientific temper is the acceptance of the premise that the method of science is the only method through which knowledge may be acquired, and the premise that all human problems can be (and should be) solved only in terms of knowledge acquired through the application of the method of science; scientific temper, therefore, requires rejection of all that is incompatible—at any given time in history—with the knowledge acquired through the application of the method of science.

The benefits of science with which humanity at large is familiar, are seldom recognized by the layman—specially in our society—to be the result of a simple, systematic, well-defined, and objective approach: the application of the method of science to the solution of problems and to the discovery of truth. The method of science stands in direct contradiction to the ways of religion, dogma and faith which are based on the premise that truth can only be revealed; revelation was, in fact, the only approach available for solving problems and discovering truth till a few centuries back. The scientific method—the seeds of which were sown by Roger Bacon in the thirteenth century—rejects revelation as a means of discovering truth and substitutes it by the technique of observation, followed by careful experimentation and logical deduction. Thus, in contrast to revelation which is highly *personal* and subjective, the method of science is universal and objective.

It is only logical that once science has permeated virtually every level of human existence, the method of science must become a way of life, and science must be conceived not merely as a body of facts but also as a way of thinking. An unqualified acceptance of the latter role of science is 'the scientific temper'; it is a corollary of the acceptance of the method of science as the only way of discovering truth.

Let us come back to the case of Shri Satya Sai Baba. Science tells us that human problems—for the solution of which people often go to godmen such as Satya Sai Baba—can be solved only in terms of man's own knowledge and experience and not by invoking the supernatural. It is undoubtedly true—and no one realizes this better than those who possess scientific temper—that there are great chasms and unfilled gaps in man's knowledge of himself and of nature, but these gaps can be filled only through application of the method of science, that is, by objective and rational inquiry.

History—from Copernicus and Galileo, through Darwin and Huxley, to the modern space age—tells us that whenever a conflict has arisen between science-based knowledge and supernatural belief, it is the supernatural belief that has been shown to be untenable. Whenever and wherever—in our country or elsewhere—claims to supernatural powers have been put to test, the claims have proven to be false. There are reasons, therefore, for being suspicious of godmen like Satya Sai Baba who claim to be possessed of unproven and untested abilities, powers or gifts which defy all reason and sensibility and which stand in direct contradiction to knowledge gained through the method of science. It seems likely that the miracles they perform and which they attribute to their supernatural powers and on which their following is largely based, are no more than practised trickery. Against this background if someone claims to possess supernatural gifts which seemingly defy all known laws of science—laws which have stood the test of time—the onus of proving that he truly possesses these gifts lies with him. Until an irrevocable proof of such powers is obtained through rigorous and open, scientifically conducted tests and experiments, attempts by an individual or a group to build-up a following by the demonstration of such gifts to credulous audiences, must be considered highly undesirable, even criminal.

The validity of claims to miracles that Satya Sai Baba makes can easily be checked through proper scientific investigation. For example, his claim to materialize objects can be checked by an adequate search of his person and apparel immediately before he performs such materialization. He, however, has consistently refused all such investigations. It is now well-known that several distinguished and responsible persons including Prof. Kavur of Ceylon and Prof. Narasimaiah, Vice-Chancellor of the Bangalore University, have written to him politely and in a spirit of inquiry, asking him to submit himself to such a test. Those who continue to believe in him in the wake of such refusals cannot be truly considered to have scientific temper, especially when it is known that everything that he can do which he calls a miracle, can be done by several magicians in the country just as effortlessly and probably more elegantly than Satya Sai Baba does them; moreover, there are many 'miracles' that some magicians can perform that the Sai Baba cannot. Indeed, if scientific temper was widely prevalent amongst our people—specially those holding positions of some power and influence—Shri Satya Sai Baba would have been probably by now in prison for engaging in activities which, by all account, tantamount to deliberate cheating and misleading people. Would not we, for example, penalize a person who insists on marketing a drug while *refusing* to allow it to be scientifically tested?

The example given above of Satya Sai Baba would show that while modern science and technology are now accepted everywhere as an integral part of one's everyday living, few ever stop to think, particularly in our country, about the obligations which rest on the users of the benefits of science and technology. The most important of these obligations is, indeed, understanding the method of science and the development of scientific temper. Let us now analyse in greater depth the need for scientific temper, then define some of the problems in its inculcation amongst the people, and finally see if we can arrive at some practical suggestions that may help in the development of scientific temper.

III

It may be perhaps argued that scientific temper is necessary only for those who practise science and that such an outlook is of little use to a person who is normally concerned only with the utilization of the fruits of science and technology and not with science and technology *per se*. There could be no greater fallacy. Today, scientific answers are either available—or there are reasons to believe (from the trend of modern researches in chemistry, biology, physics, astronomy and space science) that such answers will be found in the future—to the major questions that humanity has been asking itself since man came to be endowed with intelligence. In fact, by providing answers—or opening avenues for the acquisition of such answers—to common questions such as those pertaining to atmospheric or cosmological phenomenon (e.g., rain, thunder, day and night, and eclipses), or to the more sophisticated questions such as the nature of the universe, the origin of life, the mechanism of heredity, the cause of disease, and the basis of the various physiological processes including those which control behaviour and brain function, science has established the validity of its method and has thus given a valid technique to humanity for solving its problems. Attempts to provide answers to these questions in the past were based on the teachings of religion, dogma, faith, custom, convention and tradition. Science has proved to be right in every case where its results and conclusions have differed from traditional teaching. For example, Darwin's theory of evolution propounded a little over a hundred years ago was vehemently contested by the Church and its followers as it went against its teachings, but is today accepted almost universally.

In view of such astounding successes of science, particularly where its results came in conflict with religion, dogma or the like, it can be said with confidence that the scientific temper, which is based on an understanding and acceptance of the method of science, alone can help arrive at a *rational* solution, in terms of human resources and knowledge, of various problems which face man as an individual or as a part of the social community. Indeed, scientific temper is today a prerequisite for proper appreciation and pursuance of the ideals of liberty, freedom and self-reliance and, in the modern setting, can be a strong secular force.

The presence of scientific temper amongst the masses is necessary for developing in the people the desire to understand natural phenomenon and to acquire new knowledge, and the will and motivation to take steps that would ensure that the above desire is fulfilled. The history of the progress of mankind is a history of new knowledge acquired and natural phenomena understood.

Unfortunately, till some 50 years ago, this knowledge and understanding was confined to a small number of people—to a class which used this knowledge and understanding to exploit the vast majority for the gain of the individual members of the class. Scientific temper amongst our people would act as a strong deterrent to such exploitation.

Scientific temper could lead to a better appreciation of the impact of science and technology on society, and thereby provide the kind of atmosphere needed for science and technology to flourish and for their importance to be duly recognized. Much of science and technology done today requires expenditure of public funds directly or indirectly, and for those funds to be utilized for the benefit of most and not just a select few, it is imperative that scientific temper permeates the thinking of the masses. Scientific temper is, therefore, essential to ensure application of science and technology exclusively for the benefit (and *not* the destruction) of mankind at large.

IV

Perhaps the greatest stumbling block in developing scientific temper is the widespread belief in superstition and in the supernatural as a result of an uncritical acceptance of the dictates of religion, dogma, faith, customs, convention and tradition. The highly educated—including those trained in science—are no exception in this respect in our country. For example, Shri Satya Sai Baba counts many "scientists and intellectuals" amongst his followers. One really wonders if such people, in spite of their scientific education and often above-average intellect, have not forfeited the right to be called scientists or intellectuals. They are clearly devoid of scientific temper. Today, possession of scientific temper (much more than possession of a degree in science or of some other high educational qualification) must be considered a prerequisite for one to be called a scientist or an intellectual. This dichotomy amongst our educated should cause much greater concern than the belief in superstition or supernatural on the part of persons who have had no access to the fund of modern knowledge. Illiteracy and lack of *proper* education amongst the masses, of course, accentuate the problem. Emphasis should be laid here on the word 'proper' when used in the context of education; a large number of existing textbooks in this country, instead of propagating rational and scientific thinking, do just the opposite. The bulk of our school and college students cannot discriminate between what is legend and what is history. And it is not uncommon to find a highly educated person, even a Ph.D. in science, telling you on the basis of what he learnt in his text books, the exact date of the rule of Rama in the country! Lastly, one faces the lack of a policy and of an adequate machinery for dissemination of scientific information.

It should be emphasized that the problems we have to deal with have been with us for ages and are as much a part of the structure and fabric of our society as we would like, instead, the scientific temper to be. Indeed, all through human history, some people have ascribed to themselves supernatural powers and used these 'powers' to exploit vulnerable masses for material gain, satisfaction of personal vanity or attainment of prestige and influence, and this has happened not only in India but everywhere in the world. However, Indian godmen seem to flourish today better than those from other countries. This situation is partly a result of the use of modern public relation techniques by these godmen, and partly due to the intrinsic pre-disposition of most of our countrymen towards acceptance of the supernatural—a consequence of ignorance and of the hold of tradition, custom, convention and religious dogma on our society.

One fundamental difference between many countries of the West—socialist or otherwise—and India is that the *scientists and intellectuals* of those countries are largely emancipated from such beliefs while in our country they continue very largely to be victims of these beliefs. In the socialist countries, even the masses are largely emancipated in this respect, and I have no doubt in my mind that this is in no small measure correlated with their success in reducing exploitation of men by other men for personal or class gain.

It is of the utmost importance to make the teaching of science compulsory through schools, and to stress its *principles and methods* while teaching it. Science teaching must also emphasize what the role of science should be in everyday life, for example, in an Indian village. This would involve a complete revision of the curricula and the syllabi, the development of proper textbooks with adequate provision for frequent revision, a programme of training of teachers to teach science, and creation of adequate facilities in schools for such teaching. And all this must be done on a mass scale and not just for a select few as is done now. For this to happen, I see no other solution but to nationalize primary and secondary education. I consider this step entirely necessary—with no alternative—if we wish to ensure that scientific temper will be a way of life with our people at the turn of this country. With no central control, education (specially science education) is now the prerogative of the privileged, one of its major objectives being to exploit others, rather than inculcate scientific temper. It is heartening in this respect to note that the NCERT (National Council for Educational Research and Training) has recently undertaken to revise and reframe the curricula and the syllabi for primary and secondary education—and have the textbooks written—in a way that would help development of scientific temper amongst, specially, the rural children.

Efforts to improve science teaching should be supplemented by efforts to popularise science. Competent scientists should devote time to write articles in newspapers and help build a system and a tradition, which would ensure that science reporting in newspapers is true and interesting instead of being sensational and prosaic as it mostly is at present. As many popular science journals as possible should be started, specially in regional languages.

While there has been an increasing awareness of the possible role of radio and television in popularizing science, relatively little is being actually done to use these important media of communication to develop scientific temper in the people in an organized manner. This lacuna could perhaps be removed by the formation of an advisory committee on science for every radio and television station. These committees should consist of scientists and other intellectuals who have a strong and unambiguous commitment to science, who possess scientific temper, and who have not expressed any outright obscurantist beliefs that may be contrary to science in their private or official life. (Let us remember that *most* scientists of our country—including those occupying high positions—would not satisfy these basic criteria.) It would also be necessary to establish a code for scientific programmes on the radio and the television; this code would obviously have to be different from the one which exists at present for talks on the All India Radio.

Our scientific institutions and organizations should be charged specifically with the responsibility of taking science to the people, in addition to their other responsibilities such as teaching or research. I should like to see this objective included in the charter of every academic or research organization and institution in the country; any assessment of their achievements must take their performance in this sphere into consideration. The situation as of now is actually to the contrary. I have heard scientists occupying important positions in our country, say that they and their colleagues should be concerned only with teaching, research etc., as the case may be, and that their involvement in any other activity is not only *not* to be expected but would be undesirable.

I do not believe there has been a single instance since Independence when our scientists have taken an organized or collective stand on any social, political, economic or similar question of vital interest to the nation; they have not done so even for major *scientific* questions! Our so-called scientists and 'august' scientific bodies such as the Indian National Science Academy did not, for example, utter a word at the time of the *Astagraha*; the salutary exception was Jawaharlal Nehru! Some individual scientists have been vocal, but their number is infinitesimally small in relation to the size of our scientific and technological manpower, and they have often had to pay a price for raising their voice above the retrogressive and obscurantist ideas of the vast majority. Indeed, very few of our scientists have felt truly concerned about the tremendous amount of anti-scientific activity which goes on around us, be it by men who pose themselves as gods, or by astrologers and palmists, or by those who pretend to cure people by methods which are as antagonistic to science as dictatorship is to democracy.

It would be necessary to have our press seriously committed to science. The interest of the press is, at present, largely confined to sensational news in science; very little attention is paid to the accuracy or the scientific validity of the news item. It is difficult to have press representatives cover even outstanding scientific talks in the country containing news for which they would be otherwise willing to pay gladly if it came through Reuters or Associated Press from abroad! I could cite many instances of this type from my own institution. The reason for this, perhaps, is that science writing or reporting is not yet regarded a worthwhile career in this country, unlike in the West.

I see no reason why at least one major University in the country should not run a high-powered, properly conceived and conducted, one-year diploma course on science reporting. This, of course, will not be the solution, as most of the outstanding science writers will probably continue to come from outside such a course, but at least general standards will improve and the process of focusing the attention of people on science will be aided.

State governments should be persuaded to establish at least one science museum and one museum of natural history in every State. If such museums are established, school and college students should be encouraged to visit those museums, and as much facilities as possible should be provided for such visits by their respective institutions. These museums should stress the principles of science, the joy of scientific discovery, the substitution of the irrational with rational thought, and the importance of logical and objective thinking.

Another way in which the educated could help in the creation of scientific temper is through the establishment of associations formed with the express objectives of (a) spreading consciousness of the aims, methods and values of science; (b) removing belief in the supernatural and in superstitions; and (c) removing dichotomy amongst scientists themselves.

Lastly, we think the time has come when every scientific worker—and we include students of science and technology in this category—must do a little bit of introspective thinking as to how he may emancipate *himself* from all that is anti-science. The basic methodology one learns when one studies science and technology can lead him to only one set of logical conclusions: to renounce faith in revelation and to apply the scientific method to his everyday living. There, indeed, cannot be any substitute to the exemplary behaviour of scientists themselves as an aid to the propagation of scientific outlook, now proposed as a duty in Article 21A of the 42nd Constitution Amendment Act. It is, of course, to be hoped that the Government too will, if the bill is passed, discharge its corresponding responsibilities in this regard. The demon of Shri Satya Sai Baba cannot be fought by scientists and intellectuals alone; it would need cooperation, support and action by the State as well. And Jawaharlal Nehru's heritage calls upon us to fight such demons with all our strength.

> Our age is a different one; it is an age of disillusion, of doubt and uncertainty and questioning. We can no longer accept many of the ancient beliefs and customs; we have no more faith in them, in Asia or in Europe or America. So we search for new ways, question each other and debate and quarrel and evolve any number of 'ism' and philosophies. As in the day of Socrates, we live in an age of questioning, but that questioning is not confined to a city like Athens; it is worldwide.
>
> -Jawaharlal Nehru

IX

THE PHENOMENON OF MAHARISHI MAHESH YOGI*

P M Bhargava

Maharishi Mahesh Yogi, the originator of the technique of Transcendental Meditation (TM), now claims to have over one-and-a-half million followers—over half of them in the United States alone. He believes that his technique led to the dawn of the *Age of Enlightenment* three years ago and that, when 1% of the world population begins to practice TM, things would become all right for man everywhere, and all his problems at all levels from the individual to the state, would be solved. He would, then, have created an *Ideal Society*. Mahesh Yogi's followers call the principle underlying the above-mentioned phenomenon of transforming the whole of the society through a small fraction of its members, as *The Maharishi Effect*. The Maharishi is also the head of the "World Government for the Age of Enlightenment".

The above recipe is such that—irrespective of one's personal beliefs—one would like it to work, specially when there are other dividends promised on the way! For example, Mahesh Yogi is now attempting to show that one can, through TM, also perform the *Siddhis*. "The *Siddhis* are

* This article appeared in *New Quest*, March-April 1978, pp. 79-80.

remarkable performances described in the Yoga-Sutra of Patanjali, such as knowledge of the past and future, becoming invisible, and passage through the skies or flying", says Mahesh Yogi. Who wouldn't like to be able to fly—all by himself, without using a machine? It is, therefore, timely to have a closer look at the man that Mahesh Yogi is, the manner of his operation and, of course, his claims. Here is such an analysis, partly fact and party interpretation, based partly on personal experience.

Traversing the Knife's Edge Between Science and the Vedas

Realising that we are living in the age of science, and learning from the failures (or, at most, limited successes) in recent years of others of his genre who ignored science, Mahesh Yogi has done his best to woo science. What is specially impressive is the manner in which this has been done. The main thrust of his present preaching has been that all he has said in the past has since been proven by scientific experiments, and that there are perfectly valid scientific explanations for all that he says now.

One is also impressed by the scale on which he operates in regard to his flirtations with science—as, indeed, in everything he does. At his well-endowed Maharishi European Research University (MERU) at Seelisberg in Switzerland, the stress is on science. The chancellor is a physicist; it, of course, does not matter how good Mahesh Yogi's scientists are because he knows that in the eyes of *most* of his followers, the quality would be unimportant.

When he hobnobs with scientists, he takes up the role of the oracle. He has pronounced that all that science has done in the past, and is doing now, is to validate the Vedic truths and to provide explanations for them so that Reason is satisfied. He expects his scientists at MERU and others around the world whom he is willing to provide handsome financial support, to obtain evidence in support of his thesis. Therefore, neither he nor his scientists tell their followers and listeners that—leave aside the Vedic period thousands of years ago—even at the beginning of this century, the structure of DNA and the genetic code which have led us to understand the basis of similarities and dissimilarities in the living world, were completely unknown; and so were the principles underlying the laser, the radio and the transistor. And there is no counterpart in the Vedic literature of the theory of relativity or of the theory of evolution and, indeed, very little overleap between modern scientific knowledge and what is written in the Vedas. Mahesh Yogi and his scientists do not tell their audience that the concepts of creation, soul and rebirth which are central to Vedic beliefs, are utterly incompatible with modern science. There is little similarity between Vedic cosmology and the science-based cosmology of today. One could, of course, argue—as probably Mahesh Yogi would—that where modern science is at variance with the Vedic truths, science has not yet found the ultimate answer. But then, he would forget to tell you that whenever there has been a conflict in the past between science on one side and a dogma-based belief on the other, it is always science that has been vindicated!

An important attribute of science is its ability to make testable predictions. Thus the existence of elements, gallium, scandium, germanium and astatine; of the fundamental particles, neutrino and omega minus; of the planets, Neptune and Pluto; and of the species, Latimeria, Pithecanthropus and Oreopithecus, was predicted. Furthermore, the new element, particle, planet or species when discovered, was found to have very much the same properties as were predicted for it. For example, for Gallium, Mendeleef said years before it was discovered that it would melt with 'the heat of hand'; its melting point was later found to be about 30°C. Mahesh Yogi and his scientists, while discoursing on the compatibility of the Vedas and science, do not ever state that there is no known case of such a prediction being made in the Vedas and subsequently verified.

Indeed, the basic method of science as first enunciated by Roger Bacon in the thirteenth century and then refined to its present form through the contributions of persons such as Francis Bacon, Descartes, Newton, Darwin, Marx and Engels, Einstein, Bernal, Blackett and Fermi (to name a few), has no counterpart whatsoever in the Vedas. On the contrary, the content of the Vedas, taken as a whole, goes contrary to many attributes of science, of its method and of the knowledge gained through it. For example unlike the Vedic concept of objective truth, scientific truths are truths by consensus (amongst experts, of course) and are not unchangeable or unmodifiable. The right to question is fundamental to science, and denies credibility to people (like Mahesh Yogi) who believe that whatever they say must be accepted on faith. Science is progressive and forward-looking, and the total fund of scientific knowledge increases with time. Science does not claim to provide answers to all legitimate questions that may be asked at any time, and it often progresses by disproving. Further, science is secular and objective, and scientific observations are verifiable, repeatable and independent of what one may personally want to happen. If you mix pure sulphuric acid and water in equal proportion, the temperature of the solution will go up, irrespective of what you may wish to happen. And a time reaction set for 19+1 seconds shall take that much time even if all the Mahesh Yogis in the world, with their knowledge of the Shastric Siddhis and the rest, wish it to be otherwise! Unlike the Vedas, science does not seek an explanation of the unknown in terms of another unknown. In these respects, and many other, science and the Vedas stand opposed to each other rather than supporting each other as Mahesh Yogi and his scientists have been propagating. But these arguments are never presented by them. The caste system, one of the greatest scourges of India, was sanctified by our ancient literature, but you would never hear Mahesh Yogi discussing this issue. (This, of course, is not to denigrate the considerable amount of wisdom contained in the Vedas; it is only to make the point that not all that is said in the Vedas has been found to be scientifically tenable.)

Nevertheless, Mahesh Yogi himself as well as his followers project him—quite subtly, of course—as a superman who knows all there is to be known. He is therefore, a man apart, to be treated and revered differently from others. It is propagated by him and those of his clan, not only that all knowledge was written in the Vedas thousands of years ago, but also that he alone today understands it all; moreover, he has also discovered the physiological source of it, and knows how to communicate it to—and share it with—others. On 10th January, 1977, he wrote: "Worldwide, the Transcendental Meditation Programme is taught in over 100 countries; there are now 1,500,000 Citizens of the Age of Enlightenment, 12,000 Teachers of the Age of Enlightenment, 2,500 Governors of the Age of Enlightenment centres." This 'achievement', he believes, entitles him to be treated like a god (which treatment he both expects and loves; you only have to look at the publicity material to be convinced of this!) by his followers. Yet he would refute the charge, if it is made, that he thinks he is a god! Here, again, in sheer sophistication of technique, he scores over all the others of his kind.

The Scientific Claims

The scientific claims of Mahesh Yogi and his scientists have been largely refuted (e.g., Pagano et al., *Science*, 1976, **191**, 308, Michaels et al., *Science*, 1976, **192**, 1242). Few, if any, of the papers of his scientists during the last five years, claiming miraculous powers for TM or for his more recent interest, the Siddhis, have been published in a good journal (say, a journal indexed by *Current Contents*). Their papers now seem to appear mostly, if not exclusively, in the house journal of Mahesh Yogi's own 'University', the MERU. This journal is not indexed in *Current Contents* and I wonder how many copies of it are bought by libraries of first–rate scientific institutions around the world. Of late, his publicity material—which is as expensively brought out as it is untruthful—refers to a book, *Scientific Research on the Transcendental Meditation Programme: Collected Papers* (MERU Press, 1976). I wonder what proportion of good scientists around the world have seen the book—even in a bookshop!

The credibility of Mahesh Yogi's scientists in the world's scientific community today seems to be low. An incident in which I was involved would illustrate this point. In 1975, I participated in an international symposium on contraception at which one of Mahesh Yogi's scientists, Dr Elliot D. Abravanel, from the Yogi's International University was also one of the 30-odd invited participants. We were later told he had invited himself to the symposium. What was worse, he presented a shockingly bad paper. The paper was, in fact, scientifically so poor that it led most of the other participants—all reputed biologists—to tell the organizers of the symposium that if that paper was included in the proceedings they would not allow their own papers to be published under the same cover! The proceedings of the symposium have now come out without the paper of Dr Abravanel (K.R. Laumas (ed), *Developments in Contraceptive Technology*, Ankur Publishing House, New Delhi 110 016).

I might add an epilogue to this incident which the readers may find interesting. In his talk, presented without any data, Dr Abravanel repeatedly tried to make the point, using involved arguments that no one understood, that the technique of TM could be used for family planning. At the end of the talk. I asked the Chairman, Dr Shelden Segal, Vice President of the Population Council, New York, for permission to ask the first question. I inquired, "Will the learned speaker tell us if, for family planning, he recommends TM before or after?". Before Dr Abravanel could deal with the question, the audience replied in remarkable unison seldom witnessed at scientific meetings, 'instead!'

Indeed, very much more—and better—proof will be needed to validate the scientific claims of Mahesh Yogi and his band of 'scientists'. Most of the good scientists around the world today would, I imagine, consider Mahesh Yogi and his organization as a major fraud committed on gullible people. There are, of course, several established scientists, like Professor E C G Sudarshan, a physicist at the University of Texas at Austin, USA, and Professor of Physics, Indian Institute of Science, Bangalore, India, who are known to be his ardent supporters, However, the support of such scientists to Mahesh Yogi does not seem to increase the credibility of his claims and pronouncements. For example, at a series of recent meetings, on which I shall comment in greater detail later, organized by Mahesh Yogi in India, the Yogi said that about 3,000 persons had been initiated into the art of flying as described in Patanjali's Yoga-Sutra, and had now become "flying humans". "One can rise from the ground gracefully and without effort by spending a little time each day in Transcendental Meditation: the body just takes off and there are no complications", he said. Dr Sudarshan, who was "Permanent Director" of these conferences, endorsed the above claims of Mahesh Yogi. Dr Sudarshan, however, did not respond to any of the several requests made by responsible people for a live demonstration of the act of flying under experimentally controlled and verifiable conditions. There was plenty of time and opportunity for him to accede to these requests and, as a scientist, one would have expected him to do so. His not doing so while publicly supporting Mahesh Yogi's claims in this regard, did not help gain support for Mahesh Yogi among the intellectuals. If anything, it made people more skeptical.

One should, of course, like to understand why even a few well-known scientists who have produced good work in the past have been attracted towards Mahesh Yogi. One reason could be the exceedingly good treatment that is meted out to them by Mahesh Yogi. They immediately come to occupy high positions in the hierarchy of his fabulous empire. They now have enormous power and influence, the like of which they would have otherwise probably never known. And they have access to virtually unlimited resources as long as they are prepared to play second fiddle to Mahesh Yogi and at least make it appear to the public that they accept him as their master whose word is law for them. They receive enormous world-wide publicity, including in countries where very little science is done and where they would probably have remained unknown even if they were to win a Nobel Prize. They find they now have a following of an admiring million or two who will never question whatever they say as long as they remain high up in the Maharishi's hierarchy; they know they would never achieve such an unquestioning following as a scientist alone. Their incursion into Mahesh Yogi's fold provides them with a short cut to "success" which they would have never achieved through straight science. This lure is probably too much for some to resist. And intelligent as they are (otherwise Mahesh Yogi wouldn't have them), they have no difficulty in finding on alibi for their association with Mahesh Yogi. It is an alliance which suits both Mahesh Yogi and them perfectly. Indeed one of these scientists may succeed Mahesh Yogi, and if the prospects are of an heirship to Mahesh Yogi's empire, the lure would be even more difficult to resist! One, no doubt, finds the same spectrum of human weaknesses among good scientists as among others.

The Ungodly Acts of the God

Mahesh Yogi recently organized a series of meetings in some ten towns in India on science and the Vedas. The twin themes of these meeting were "Consciousness: the Field of All Possibilities", and "Knowledge is Structured to Consciousness". (Like many others, I have wondered what these phrases mean.) In the newspaper advertisements (for example, in the Hindu of 3rd June 1977) and in an 8-page, expensively produced pamphlet brought out for these meetings, the name of Dr Raja Ramanna, then Director of the Bhabha Atomic Research Center, Bombay, and a distinguished physicist who was partly responsible for development of the expertise that lead to the explosion of the first Indian nuclear device in Pokhran in May 1975, appeared in the list of sponsors of the meeting. I am giving below an extract from the reply I received from Dr Ramanna in response to my request to him to confirm that his name was being used by Mahesh Yogi with his permission and knowledge. Dr Ramanna wrote:

> I am afraid I knew nothing about the All India Conference on Veda and Science, except that Professor Sudarshan mentioned it to me while he passed through Bombay and suggested I attend one of the meetings. Unfortunately, I could not attend any of the meetings due to other commitments. While being curious about the para-physical phenomena, though I do not believe in them, at no stage did I agree to be a sponsor.

And this would not be the first time that Mahesh Yogi had deceitfully exploited well-known names in science for his publicity. The well-known periodical, *Science*, reported in its issue of 28th March 1975. "The backbone of Maharishi's movement is made up of young people, most of them volunteers who are ardently devoted to the guru and his principles. MU (Maharishi's International University) tries to give the impression that it has the endorsement of great minds in scholarship and science, whose names are scattered about the catalog. But such is not quite the case. Chemist and Nobel Prize winner, Melvin Calvin of the University of California at Berkeley, says he addressed one of the SCI [Science of Creative Intelligence, a term coined by Mahesh Yogi (this explanation is by the author)] symposia, but he considers his name in the catalogue as coming "perilously close to false advertising".

One of the conferences in the series mentioned above was held at Hyderabad on 7th August 1977, and I was invited to speak at it. Mahesh Yogi and his followers must have, later, regretted their persistence in asking me to speak, even though I had made it quite clear that if I did speak at the Conference it will only be to express skepticism of his claims. It seems they did have second thought about inviting me as the original invitation, extended more than two weeks earlier, was confirmed only 24 hours before the Conference. However, the omniscience of the Yogi and his organizers failed to recognize the streak of piggishness in me which made me say at the meeting exactly what I felt, even though I thought I didn't have ten supporters out of several thousand who were present, and for the six hours preceding my talk there had been respectful homage to Mahesh Yogi by speaker after overwhelmed speaker. Much worse, they confused this streak for courage; at the end of my talk, Dr Sudarshan in an unscheduled speech—apparently made to assuage Mahesh Yogi whose response to my talk, as I was told later by many in the audience, was quite human—admired my exceptional courage (courage for what, I thought: for criticizing, and challenging a god?). But let us get back to the narrative.

When at a few minutes past eleven in the morning, the Governor of Andhra Pradesh, Mrs Sharada Mukherjee—a charming, modest and dignified lady who does more than justice to her job—arrived on the stage and was introduced to Mahesh Yogi, neither he nor anyone else on the stage, including the so-called Vedic pundits sitting on the dais, got up from their seats. Nor did they get up when she left after an hour, after inaugurating the conference. Some of us wondered: what happened to the Vedic tradition about which Mahesh Yogi had just talked so glibly, which tradition certainly prescribes a show of respect to women if not to the head of State. And we did learn he knew how to rise from his seat to greet, as he did so quite elegantly later, in honour of a Vedic pandit. Besides me, there were four other invited speakers, including a former Vice-Chancellor of Osmania University, a former Principal of a well-known University College, a distinguished psychiatrist, and a senior civil servant belonging to the Indian Administration Service. We spent more than nine hours at the meeting but the Yogi didn't come down from his pedestal even to say 'hello' to his invited guests. He perhaps expected them to go and bow before him or to touch his feet or—more likely—to put their heads on the path his feet had trodden as many others in the audience actually did! And to make sure that he was not rendered impure by the crowd, it was announced that no one might leave his seat until he had departed—in right royal style! It did seem strange to me as I was not used to such protocols at the numerous scientific meetings I have attended around the world—and this meeting was announced as a scientific meeting.

Why Such a Large Following?

One may now ask, what is there in Mahesh Yogi and his teachings which impresses people, at least some people. A precise answer to this question would probably require a considerable sociological research input—in fact, that may be one use to which he could be put—but some reasons are obvious. For example, he uses meaningless, bombastic phrases which no one, including probably Mahesh Yogi himself, understands: just enough mysticism to fool people but not enough sense to make them ask a specific question! For example:

> "Consciousness: The field of all possibilities."
> "Knowledge is structured in consciousness."
> "Flood of knowledge through science will verify Vedic Truths."
> "Every accomplished scientist in India is a Vedic. The Veda content flows through his veins".
> "The time has come for man to become *perfect*."
> "The law of least action."

The above sentences and phrases are reported here verbatim from his speeches and interjections on 7th August 1977 at Hyderabad. Most of what he said was a collection of similar phrases: nothing concrete, nothing definitive, all wooly.

His disciples are almost as well trained in the art of grammatically correct gibberish. As an example, look at the following excerpt from the talk of Dr. David Orme Johnson, Vice-Chancellor of MERU, at Jammu in India on 16 January 1977:

> The simplest form of awareness is the 'home of all the laws of nature', parallel to the simplest form of matter and energy, the vacuum state of the quantum field. We know from quantum physics that all forms of matter and

> energy are present in their potential form in the vacuum state, although nothing is manifest. It is the null set, able to become anything not yet precipitating into anything. In the same way consciousness is the unmanifest source of all the laws of nature. Because pure consciousness is the home of all the laws of nature, it is possible to make all activity evolutionary. When action is profound on the basis of stabilized pure awareness, it is completely evolutionary for the individual and for his environment. This is the basis of our global undertaking to create an ideal society. The development of the field of pure knowledge in the awareness results from the stabilization of pure awareness through the regular practice of the Transcendental Meditation technique.

Impressed? Perhaps a computer, given the rules of grammar but otherwise asked to choose randomly from the OED could have done no worse.

On listening to Mahesh Yogi on 7th August 1977 at Hyderabad, I wondered if he ever pondered before he said anything. All the un-sense seemed to come to him naturally! One of the phrases he repeatedly used in his many talks and interpretations during the day was "pure knowledge", to acquire which, he said, should be our ultimate aim. In reply to the query in my talk about this phrase ("What is *impure* knowledge, if there exists *pure* knowledge?", I had asked), the Chancellor of MERU, Dr. Lawrence H. Domash said in his long-winded, unscheduled speech after my talk that the ultimate knowledge you get through TM was pure knowledge! Of course, he did not say it all so plainly: that wouldn't have been in keeping with the Mahesh Yogi style. He used even more meaningless phrases to explain the one I had enquired about. He took nearly 30 minutes and I felt genuinely apologetic to the audience for having asked the question. Incidentally, Dr. Domash also said that once you had acquired pure knowledge, there was no need to test it through experiments. He was obviously unaware of the damaging effect that the Cartesian tradition has had on French science for precisely the same reason—but then how many in the audience had even heard of Descartes? For most of them probably, Dr. Domash's answer was satisfactory, and I was given no opportunity to question him again.

The second obvious reason of Mahesh Yogi's success would seem to be his exquisite public relations technique. Every publication that he brings out, be it a pamphlet or a book, is superbly produced and often in excellent taste, as long as you don't read what is inside. Every move he makes, every step he takes, could be an object lesson to advertising and publicity professionals. The photographs are always superb. The graphs based on his "scientific" data are meticulously and beautifully presented; I do not know of any scientist who has ever done better. And if all this is done, what does it matter if there is a scientific error here, or a scientific incompatibility there, in his data? After all, a vast majority of those would look at the graphs would be strangers to the field anyway!

The third reason seems to be his policy to avoid and evade all challenge. It is probable that I was the first person to challenge him anywhere and at any time so openly and before such a large gathering as on 7 August 1977 at Hyderabad. It could very well be that inviting me was probably the first (and the last) error of judgement of its kind he or his followers had ever made. Mahesh Yogi pronounced that one of the *Siddhis* that he had been practicing and teaching with success is the flying *Siddhi*. One of his pupils, Miss Saroj Pande who had delivered me the invitation to speak, had earlier told me that some of his pupils could now fly about in the room. And there were photographs all over (in the brochure, behind the stage, etc.) of people sitting cross-legged in the air some distance (say a few feet) above the ground. I politely challenged him in my talk to demonstrate the flying *Siddhis* in the same hall at 8.30 PM that evening at which time he had called a meeting of his meditators (I believe some four thousand turned up), and allow me to investigate his claims with the help of instruments I would bring from my laboratory. The challenge was not taken and no reasons were given. I later learnt that on this particular point, even most of his followers in the audience, as well as the press, were disappointed. His followers, of course, believed that he could do it and it would be nice to settle my doubt—which they conceded was legitimate—once for all by a demonstration. But here lay the cleverness of Mahesh Yogi. He knew that his followers would really never doubt him as far as I was concerned, I would be soon forgotten. He would just make sure that no one challenges him again!

The local and the national press completely blacked out my talk even though I am not entirely unknown to the press, and what Mahesh Yogi and his followers, and admirers said was covered by the press in considerable detail. That he knows how to manage at least an influential segment of the press is another of his "strong points".

Then he knows how to appeal, *concurrently*, to the religious sentiment in most people around the world which leads them to revere things which are ancient and connected with the mystique of religion, and to the sense of reason and objectivity arising out of the key role that science and technology play today that they cannot ignore.

And lastly, Mahesh Yogi must be one of the shrewdest businessmen around: he may even be one of the richest. (This, I would imagine, would be worthy of investigation; income-tax authorities in some countries may even benefit by it!) He charges, and heavily too, for every service rendered. And if he ever gives it free to anyone, it is no more than a free sample of medicine given to a doctor by a medical representative. It seems that TM gives to the rich for a large sum of money what the poor obtain for themselves for nothing through relaxed, sound sleep. As W.J. Cromic wrote in the 11th January 1976 issue of *San Francisco Sunday Examiner & Chronicle.*

> TM is not the only way to meditate and the flowers, incense and mantra it includes are largely hokum. TM can lower your blood pressure and reduce anxiety. So can other methods of meditation. So can a short nap twice a day. In fact, brain wave recordings of people engaged in TM show that many of them spend more than half the time asleep. A lot of people are paying $ 125 to learn to relax enough to take two 20-minute naps each day. Any method that gets people to relax for 40 minutes a day can produce the same feelings of well-being and alertness as TM.

If indeed Mahesh Yogi had something so vitally important to give to the people, why should he charge? Jesus Christ or the Buddha didn't, nor did any of the other holy men around the world who have lived through the ages and who have genuinely felt that they had something to give to people, something that was not material but yet could contribute to their happiness and contentment. I asked this question of Dr. Ram Sahay, former Vice Chancellor of the University of Allahabad, who is now in the employ of Mahesh Yogi, holding an important position in his hierarchy. Dr. Sahay snapped, "Why shouldn't he charge?" and I realized I had asked the question of the wrong man. If Mahesh Yogi hadn't been charging, who would have taken such handsome care of Dr. Ram Sahay and a host of others like him?

The Lack of Success in India

Mahesh Yogi had, on 10th January 1977, according to MERC Press Publication No.G1180 from which the so far unidentified quotations given above have been taken, 868,597 followers in the United States but only 68,700 (out of whom many may be foreigners) in India. On a per unit population basis, therefore, there are about 10 times as many followers in the United States as in India. Why so, even though the Maharishi comes from India? He himself seems concerned about this; hence the fabulous display and publicity at the series of recent conferences in India I have talked about.

I do not claim to understand all the reasons for this situation but a part of the answer may lie in the nature of the Indian background. There are many more Indians than foreigners who could perhaps challenge his claims about the Vedic tradition. But most important of all, Indians are not rich and they simply wouldn't be good business proposition. Only when the foreign market is about to be saturated (as may be the situation now) would he turn towards India.

The reaction of the intellectual elite in India to the Mahesh Yogi business is well-illustrated in the reply sent by Dr. V Siddhartha of the Indian Space Research Organisation to Dr. K P Sinha, who invited him to participate in the Bangalore meeting of the recent series in India on Science and Vedas that I have talked about above. Dr. Sinha is a physicist colleague of Dr. Sudarshan's at Bangalore, and an ardent follower of Mahesh Yogi. In keeping with his policy of wooing scientists, Mahesh Yogi has made Dr. Sinha President of the Maharishi Indian Research Academy at Bangalore. Dr. Siddhartha wrote in reply:

> I must confess that I was surprised to receive your kind letter of invitation to participate in a Conference on "The Vedas and Modern Science". I was surprised because I did not know that you subscribe to the ideas of HH

Maharishi Mahesh Yogi (incidentally, what does HH stand for?). The logo on the top of (your) letter says: Knowledge is structured in Consciousness. I am afraid I do not understand this, if it is supposed to be anything more than a tautological statement. Also, I do not know much about consciousness, not being a brain biologist, and I know even less about 'knowledge', much less the limits of it.

Not Alone

Fortunately or unfortunately, depending on how one looks at it, Mahesh Yogi is not alone in what appears to be a marvelous business of exploiting gullible people who are looking for short cuts to happiness and contentment, mental and material. There is the Satya Sai Baba of Puttaparthi (a place near Bangalore) with 20 million followers, including a smattering of non-Indians. And recently I saw the following advertisement in the 7th August 1977 issue of *The Illustrated Weekly of India* for "Tele-response power" which

> ... automatically brings you anything you desire, and in 10 seconds starts to draw Riches, Love, Fine Possessions, Friends, Secret Knowledge and much more into your life? Yes, a staggering miracle has happened: A brilliant psychic researcher has discovered a secret so powerful that it is said to bring your desires to you, from the invisible world, like a blazing streak of lightning! Yes, how would you like to be able to sit in your living room, give the command for love, and instantly have your loved one appear at your side? Or give the command for money and suddenly find a big thick roll of rupees in your hand?

Glory to them *all*! It is much better that way than to have had just one on the scene.

The Cure

I can think of long-range remedies which might prevent such phenomena as Mahesh Yogi from sprouting, at least on a scale that would cause concern, but I cannot think of a short-range solution. May be if people are going to learn to fly all by themselves or find out how to become invisible through TM, we will have to think of new laws. We may even have to think of prohibiting or regulating TM as is being done the world over with experiments on genetic engineering!

Surprising, no one has yet written a science fiction book on TM, nor has it caught the attention of Hollywood or Bombay. Any takers?

X

THE METHOD OF SCIENCE AND ITS VALUE SYSTEM*

P M Bhargava

There are no secrets in science. One can do and know all that a scientist does and knows. One does not need to have any faith in him as an idividual and accept what he says without questioning it. The only thing a scientist asks of one is to base his questioning on simple, untempered and uninhibited reason, the kind of reason that characterizes a child on whom society has not yet imposed its beliefs and prejudices. In fact, science is nothing more than questions asked and questions answered, and the method of science, no more than a method of asking questions and answering them, a method that "works" and can be used to solve problems that one faces in everyday life.

Science is, indeed, far more than just physics, chemistry, biology, astronomy and mathematics. It is a question of ideas and a way of thinking; it is a culture and a philosophy of life, a philosophy which allows us to pursue truth without any prejudgment. What, then, is this attitude of mind, this culture and this philosophy of life? It turns out that all these—and other—concomitants and attributes of science

---

* This article is based on a lecture delivered at the Asiatic Society of Mumbai's 175th Anniversary Lecture Series on Science and Society. It appeared in *Social Science Probings*, September 1984, Vol.I, pp.417-435.

emanate primarily from the method that science uses to acquire knowledge. Certain characteristics of this method, which we call "The Method of Science", and of knowledge gained by the application of this method, create a value system within their framework, a value system which is rational and reasonable, which appeals to common sense, which is commensurate with knowledge and, above all which has a built-in corrective. It, therefore, becomes specially important to understand the basis of the method of science, and to have a look at the attributes of knowledge gained through this method and at the value system of science, so that we may understand the culture, the attitude of mind and the philosophy of life, which are the essence of science. This is what we shall endeavour to do here. First let us see what the method of science proposes to do.

Let us take a set of eight statements and look at the truth value of these statements:

(1) The sun rises in the west and sets in the east.
(2) If you sow Sonora wheat, you will reap Sonora wheat.
(3) Water can be made from hydrogen and oxygen.
(4) Some objects can travel faster than light.
(5) Atoms are unbreakable.
(6) There is life out in space.
(7) It is a bad omen if a black cat crosses your path.
(8) Family planning cannot solve any problem.

Out of these eight statements, in my opinion (and I hope you would agree with it), two statements (2 and 3) are right, and four (1, 5, 7 and 8) are wrong; in the case of the remaining two (4 and 6), the answer is not known—that is, we cannot say as of today whether the statement is right or wrong. Given this situation, one may now ask the following questions:

(a) How was the right answer arrived at?

(b) If we do not know the answer today, how will we be able to find the answer in the future?
(c) If you happened not to agree with my answer, how could I say with confidence that you were wrong?
(d) If you would like to verify my answer, how would you go about it?

The answer to all the above questions is: by using the method of science. The method of science allows one to obtain true and reliable answer to questions.

The Steps in the Method of Science

There are four distinct steps in the method of science: the framing of the question, framing of a hypothesis, doing of an experiment, and arriving at the answer which may be a fact or a generalization in the form of a theory or law. At each step in this sequence and in going from one step to the next, we use existing knowledge and logical reasoning.

The questions in science arise out of careful observation or careful analysis of existing knowledge: there is no third origin of a question! In fact, if you have framed your question properly, you are already on your way to finding the answer. What, then, is a well-framed question? A well-framed question is the one for finding the answer of which means are available within the framework of the method of science. Such a question should lead to a hypothesis which can then be tested by an experiment. Let us take an example.

Out of the million species that inhabit our earth, nearly seven hundred thousand, i.e. some 70% of them (all insects), have six legs—indeed, a remarkable observation, having tremendous implications. Well-framed questions arising out of this observation would be:

(a) How did all the seven hundred thousand or so species come to have six legs?
(b) Do these species have other common features?

(c) Could they have originated in nature from a common ancestor?

These are well-framed questions. The more carefully you observe, the better-framed your question would be.

The second step, the hypothesis, is an answer we may consider possible. The single most important attribute of a good hypothesis—a scientific hypothesis—is that it must be testable. A testable hypothesis is one which can be tested by an experiment or on the basis of which a testable prediction can be made. For example, a common place observation is that objects, it left unsupported in space or earth, will fall to the ground. One can make many untestable hypotheses to explain this phenomenon. One such hypotheses would be they fall to the ground because a particular friend of yours wills them to do so. Another would be: they fall to the ground because God desires them to do so. None of these hypotheses is testable. Therefore, they are not scientific. On the other hand, if you make the hypothesis that an object unsupported in space on earth falls to the ground because there exists a force of attraction between the object and the earth, you can test the hypothesis. In fact, it was starting with such a hypothesis that Newton arrived at his famous laws of gravitation.

The third step in the method of science is the experiment. The experiment must have one of the following two objectives: either to find an answer to a question, or to prove or disprove a hypothesis. An experiment which does not attempt to meet one of these two objectives is unlikely to be a good experiment. To do an experiment, one must make an inventory of all the steps in the experiment, collect all the material one needs, and carry out all the steps with the utmost possible care. More important, one must record all the observations, paying attention to the smallest detail.

And most important, one must record everything—expected or unexpected—whether one wished it to be so or not. It is in the doing of an experiment in this manner that the values of objectivity, lack of bias and the exercise of care in what one does, come into existence in science. An unexpected observation could lead to an important discovery. We may recall that penicillin was discovered by Alexander Fleming in this way—because of an entirely unexpected observation he made while trying to test quite a different hypothesis.

The last step in the method of science is the answer. The answer generally takes one of two possible forms. It could either be a scientific fact of limited applicability, or a generalization of wide applicability. Let us consider an example; the case of the two diseases, albinism and haemophilia.

Albinism is the lack of the pigment (called melanin) which gives the skin its natural colour; albinos have an unusual appearance and can be easily spotted: their skin is pinkish white, their hair has the colour of straw, and their eyes are sensitive to intense light and need to be protected. Haemophilics do not posses the ability to allow the blood to clot, so that a haemophilic would bleed to death because of a small wound which many of us would not even notice. The question is how are these diseases caused?

It has been observed that these two diseases are found in a cluster in families. It is also known that synthesis of melanin and clotting of blood are biochemical processes controlled by biological catalysts called enzymes. A reasonable hypothesis, therefore, would be that albinism and haemophilia are hereditary diseases caused on account of malfunction of an enzyme. One could do two types of experiments to verify the hypothesis. First, one could look for other albinos or haemophilics in the family of an albino or a hemophilic and determine the familial pattern, if any, in regard to the occurrence of the diseases, for, if the diseases are inherited, they must follow the laws of heredity. The second experiment would be to determine, by biochemical analysis, why the albinos cannot make the skin pigment, or why the blood of a haemophilic cannot coagulate, and then see if this biochemical deficiency is present in all haemophilics and albinos. Both these investigations were done. It was found that these diseases are inherited according to the laws of heredity propounded by Mendel more than a hundred years ago, and now well understood. It was also shown that the disease was due to the lack of a particular enzyme in each case. Therefore, beginning with an observation, through the construction of a testable hypothesis, and by doing experiments, we came to the conclusion that an inheritable enzyme deficiency can cause a disease: in fact, one of the most important conclusions ever arrived at in biology.

The Uses of the Method of Science

The method of science can be used for various purposes. Let us consider three such uses.

(a) *To find an answer to a specific question, the like of which has been answered many times before.* Here one asks the question, does the experiment, and arrives at the answer, without framing a hypothesis. One uses past experience and existing knowledge to design the experiment and to interpret the results of the experiment. For example, the question may be; find the density of a 50 paise coin. Indeed, scientists have found densities of objects many, many times before and the procedure for doing so is well-known. One may, therefore, skip the hypothesis step and go on to determine the weight and the volume of the coin (the experiment). The density would be weight divided by volume (the answer).

(b) *To find an answer to a question of which the answer is wholly or partially unknown.* In such cases, one begins with an observation or an analysis of existing knowledge (the framing of the question). A hypothesis is then formulated and experiments designed to test the hypothesis; the result of the experiments are recorded. If the results support the hypothesis, the hypothesis could be the answer. If so, more experiments are done to test the validity of the hypothesis. If the results of any of the experiments do not support the hypothesis, the hypothesis is modified and another set of experiments designed to test the new hypothesis. Once all the experiments that could be done to test the hypothesis have been done, and all of them have supported the hypothesis, one could say that the hypothesis is correct. The hypothesis, then, becomes the answer to the question initially asked. This is the sequence followed in scientific research.

(c) *To find an answer to a question under conditions where an experiment is not possible.* This is, perhaps, the most exciting application of the method of science. As an example, let us take the question: what should you do to help your brother who is suffering from typhoid, to get well as soon as possible? One may indeed construct many hypothesis as possible answers to this question. Should you go to a place of worship and pray? Should you let nature take its own course and hope for the best? Should you give him an extract prepared from spices or herbs recommended by your grandmother or a friend? Or should you seek the advice of a doctor trained in modern medicine? You obviously cannot test all the alternatives by experiments. Indeed, if you followed the method of science, you would determine which of these answers will be most compatible with existing knowledge. If such an exercise is done carefully, you would probably come to the conclusion that it is the last suggestion, that is seeking the advice of a competent doctor trained in modern medicine, that is in agreement with existing knowledge and is likely to serve best the objective of curing your sick brother as quickly as possible. (Modern medicine provides a sure remedy for typhoid.) The method of science can be applied in this manner for answering questions concerning a vast spectrum of human activity, from decision-making in daily life to ethics, politics, economics and social behaviour.

An Example of the Application of the Method of Science

Let us ask the *question*: what is the path followed by light. The simplest hypothesis would be that light travels in a straight line. To test it by an experiment, one should collect a point source of light (A), two cardboard sheets with pinholes (B and C), three stands suitable for the light-source and sheets, and a scale. If light were to travel in a straight line, and the light source A and the cardboards with pinholes B and C were kept in the order A,B,C, the light source should be visible through the hole C when A, B and C are in a straight line. Further, the light source A should not be visible through the hole C when the source and the holes (B and C) are not in a straight line. These are precisely the observations you would make if you were to do the two experiments. In other words, the results of the experiments support your hypothesis. The *answer* to your question, then, is that light travels in a straight line.

One can arrive at the same conclusion through many other experiments. Today, one can actually *see* light travelling in a straight line; think of the laser beam!

At times, of course, light may appear to travel in curved lines as in the case of fibre optics but, in fact, closer analysis would reveal that this is no exception to the law that light travels in a straight line. In fibre optics, light from the source undergoes a series of total internal reflections within the tube and, in this process, gives the appearance of deviating from the straight path. We are indeed glad that, in this case, appearances are deceptive, or else we would not be able to use fibre optics behind opaque barriers, as we now often do in surgery. You would, however, find that the above law—that light travels in a straight line—would break down if:

- a uniform medium was not used for the light to pass, for example, if cardboard B was placed in between two glass sheets; or
- we could measure and align distance with an accuracy of 0.00001 cm!

We may, therefore, also conclude that, when stating a law, it is important to define carefully the conditions under which the law may be expected to hold true.

Some Landmarks in the History of the Method of Science and its Applications

The development of the method of science begins with Roger Bacon in the thirteenth century. He was the first to recognise the importance of experiments and of direct observation, thereby laying the foundation of the method of science. We, then, have Leonardo da Vinci, Nicolaus Copernicus, Francis Bacon and Galileo Galilei in the fifteenth and sixteenth centuries, whose contributions to the method of science were monumental. With da Vinci began the 'new age' (the Renaissance) in Europe; it was the first time that science played a major role in social transformation. Copernicus' was the first major scientific discovery (that the earth revolves round the sun) that eventually toppled a widely-held non-scientific belief (that the sun revolves round the earth). Francis Bacon made the first formal statement of the method of science as we know it today: the making of experiments, the drawing of general conclusions from them, and the testing of these general conclusions and generalizations through further experiments. And, with Galileo of Galilei began the breach between science and religion; his was the first major case of victimization of an outstanding scientist by religious leaders because of the views he held. (Galileo supported Copernicus, became a victim of the Inquisition for this 'heresy' and died under house arrest.)

Rene Descartes (1596-1650) introduced the rigid concepts of mathematics and logic, and of the method of doubt and questioning, into the method of science. Issac Newton (1642-1727) first made use of the application of experimental method—the method of science—for arriving at a set of interrelated generalizations. The Royal Society in Britain and the French Royal Academy of Sciences, founded during 1662-1666, were the first *organizations* to support the idea of doing experiments. The Industrial Revolution that began around 1760 with the first large-scale establishment of factories, was the first application of the method of science to organized and efficient large-scale production.

Friedrich Wohler (1800-1882) demonstrated, by using the method of science, that substances made by living systems are not anything special; they can be made in the laboratory from non-living materials. Charles Darwin (1809-1882) applied the method of science to synthesise a large body of descriptive information. He enunciated the theory of evolution and the idea that man has evolved gradually from lower forms of life. Louis Pasteur (1822-1895), using the scientific method, demonstrated that life can be generated only from life, thereby laying a foundation of modern biology. Karl Marx (1818-1883) and Friedrich Engels (1820-1895) applied the method of science to an integrated analysis of social, political and economic problems, an analysis that led to the formulation of the first, science-based socio-politico-economic theory. Frederick Growland Hopkins (1861-1947) discovered vitamins and laid the foundation of the science of nutrition which tells us what to eat and how much of it to eat; not only he but nearly ten of his collaborators and colleagues subsequently won Nobel

Prizes! Vladmir Ilyich Lenin(1870-1924) stated the concept of planning, using the method of science, for development and for the achievement of social objectives.

With Albert Einstein (1879-1955) began the age of 'grand generalizations' in physics. J. D. Bernal (1901-1972) gave new dimensions to the method of science when he put forward his enunciation of the intimate relationship between science and society. Later on, with P.M.S. Blackett (1897-1973), he found that the method of science could aid successful conduct of major operations in war. Thus operational research, destined to play a vital role in war and in peace, came into being. Enrico Fermi ushered in the nuclear age by carrying out the first controlled nuclear chain reaction in 1942, on the tennis courts of University of Chicago. In the next decade, James Watson and Francis Crick brought about the modern biological revolution by determining the structure of the genetic material (DNA). And finally, the landing of the first man on the moon and his return to the earth in 1969—the twenty–first event in my listing here—demonstrated in a spectacular fashion the fact that science allows one to make testable predictions.

These landmarks show that the scope of the method of science has continuously increased even though the basic steps have remained unchanged.

How does Science Progress?

In science, at a given time, we accept a theory or a law when:

(1) all the observations made and experiments done until that time support the fact or theory;

(2) the new theory satisfactorily explains all that was explained;

(3) no new experiments can be conceived at that time, the results of which may not support the theory; and

(4) all predictions made on the basis of the theory upto that time have turned out to be right.

Then, and then alone, is a theory in science accepted.

An existing fact or theory gives way to a new fact or theory when the following criteria are met:

(1) New experimental evidence is obtained which is not in conformity with the existing fact or theory.
(2) The new theory satisfactorily explains all that was explained by the earlier theory and some additional observations not explained by the earlier theory.
(3) Predictions can be made on the basis of the new fact or theory which could not have been made on the basis of the earlier fact or theory.
(4) Some of these predictions have been tested and every tested prediction has turned out to be right.

Let us look at two examples.

Example 1. In the 17th century, Isaac Newton formulated his famous laws of motion and gravitation, which explained the motion of objects on our Earth and, to a more limited extent, elsewhere in the universe. He also showed that white light can be broken into lights of different colours which can be recombined to give back white light. Three centuries later—in the first half of this century—Albert Einstein formulated his famous theory of relativity which explains all that was explained by Newton's laws, *plus* the following phenomena /observations that cannot be explained by Newton's laws:

- The mass of an object depends on its speed.
- Mass and energy are inter-convertible.
- The speed of light is independent of the motion of the light source and of the observer.
- Light bends in the presence of a large gravitational force such as that of the sun.
- 'Black holes' exist in space into which light may enter but from which it will not come out.
- The 'wavelength' of light increases as a result of its

passage close to a massive object, such as a star. (Red light at one end of the visible spectrum into which white light can be broken, has the longest wavelength, while violet light at the other end of the spectrum has the shortest wavelength.)

For these reasons, Einsteinian physics replaced Newtonian physics, even though the former evolved from the latter.

Example 2. In the 19th century, John Dalton, a British scientist, formulated his famous atomic theory, according to which:

- All elements consist of 'atoms' which are indivisible.
- All the atoms of an element are identical to each other in size and weight but different from the atoms of all other elements in these respects.
- Atoms of elements combine in simple ratios to form compounds.

Current concepts of structure of matter (incorporated in the modern atomic theory) explain all that was explained by Dalton's atomic theory *plus* the following phenomena/observations which cannot be explained by Dalton's atomic theory:

- The intensity of forces which hold atoms together varies in a molecule.
- The properties of substances depend on their state: solid, liquid or gaseous.
- Changes occur in the properties of solvents when certain substances are dissolved in it. For example, water—a non-conductor of electricity—becomes a conductor when a salt is dissolved in it.
- Many substances are electrically charged—that is, they move towards the positive or the negative pole in an electrical field.

- Different elements may exist in forms which have the same atomic weight.
- Elements, when arranged in the order of increasing atomic weights (as in Mendeleev's periodic table), show a periodicity in their properties—for example, in their relative ability to combine with other elements.
- Certain elements are radioactive, but not others.

The modern atomic theory, therefore, had valid reasons to replace Dalton's atomic theory.

The Right to Question

One of the most important attributes of science is the right to question. Knowledge advances and science progresses because people exercise their right to question. However, to question existing knowledge (a fact, theory or law) without any rational basis or reason is as unscientific as never to question at all. The reason for the questioning may be a flaw found in an earlier experiment; a known observation which was earlier ignored and which can be shown to be incompatible with the earlier fact, theory or law; an alternative explanation found for the evidence on which the earlier fact, theory or law was based; or new evidence which is incompatible with the fact, theory or law. Therefore, science puts a constraint on the freedom to question while making the right to question a "fundamental" right. Let us look at two examples, one mediaeval and the other, modern.

Five hundred years ago, it was universally believed that earth was the center of the universe and the sun and the planets revolved around it. Copernicus (1473-1543) questioned this belief. His questioning satisfied the criteria stated above. He used the method of science and his creative ability to show that his doubt was justified and that the age-old belief was wrong.

It has been and still is (as the school textbooks will show) a common belief that there is one and only one point on the surface of the earth from where if you travel one mile south, then one mile east (or west) and, finally, one mile north, you would reach where you started from. This point is obviously the North Pole. A few decades ago, someone questioned this belief, and came out with the interesting finding that there are an infinite number of points lying on a series of loci from where one could do exactly the same thing: that is, go one mile south, one mile east, and one mile north and come back to where he started! (If you cannot work this out yourself, write to the authors for the answer.) This is a remarkable example of how science progresses because people exercise their right to question.

Science has no High Priests who cannot be Questioned

Another attribute of science is that it has no true high priests. No scientist can insist that whatever he says must be accepted without questioning, that is, entirely on the basis of faith in him. In fact, there has probably never been a scientist, including the most eminent ones, who has not been questioned and shown to be wrong sometime or the other. Science, therefore, does not accept the existence of godmen, that is, men who believe they have special powers that cannot be understood by other men, and whose statements and actions must, therefore, be accepted by others without questioning. Science is, therefore, democratic, non-exploitative and non-dogmatic.

Science Often Progresses by Disproving

In fact, discovery in science often disproves an earlier scientific belief, that is, a part or whole of an existing scientific fact, theory or law. For example, at least two Nobel Prizes were awarded for discoveries which were later partially disproved:

(a) In 1927, Heinrich Wieland received the Nobel Prize for chemistry for discovering the structure of cholic acid, the parent substance from which are derived a large number of important chemical compounds, such as cholesterol (an excess of which in the body is correlated with heart disease). A part of this structure was proved to be wrong soon afterwards.

(b) In 1959, two American biochemists, S. Ochoa and A. Kornberg, received the Nobel Prize for physiology and medicine for the discovery of enzymes (biological catalysts) which carry out the synthesis of nucleic acids—the chemical substances responsible for heredity. Later, it turned out that neither of the two enzymes discovered by Ochoa and Kornberg was responsible for the synthesis of nucleic acids in living systems.

Let us look at a few other examples.

Einstein's theory of relativity enunciated in 1905 disproved several beliefs held by scientists. For example, until then, the following statements were generally held true by all scientists: (i) the mass of an object does not depend on the speed at which the object is travelling; (ii) mass and energy are not inter-convertible; and (iii) speed of light in free space depends on the motion of the light source and the observer. Today, we know that these statements are incorrect.

Proteins are constituents of our diet which are essential for us to survive, grow and be healthy. It is, therefore, important to know how much protein we need to consume everyday. Estimates of the daily protein requirements of an average, normal adult male have, however, changed drastically over the years:

Year	**Estimated Daily Protein Requirements**
1881	145 grams
1936	65 grams
1965	46 grams
1973	37 grams

Obviously, not all the above values could be correct!

Till recently, it was widely believed that the major nutritional problem in the developing countries such as India, was the problem of protein deficiency—that is, people did not eat enough protein. Recent work has disproved this theory and shown that the major problem is a deficiency of protein as well as the other, major energy-producing constituents of our diet; in other words, there is just not enough food for the people!

Estimates in the past of such a simple parameter as the density of air have differed very significantly from the value accepted today:

Year	Experimenter	Density (specific gravity) of Air
1600	Galileo	0.0025
1642	Descartes	0.0067
	Mersenne	0.0007
1659	——	0.0011
Today	——	0.0013

It is important to recognise that in every case cited here (and in all other similar cases), the earlier fact or theory was disproved only in the sense that it was modified—that is, altered partially and not replaced entirely—by the new discovery. The new discovery would probably have not been made—at least at the time it was made—if the earlier fact or theory had not existed before. In every case, the earlier fact or theory provided the starting point for the new discovery.

Science is not Dogmatic or Unreasonably Insistent

Science is not orthodox or conservative. Scientific truths are arrived at by an agreement of opinion—that is, by a consensus—based on the method of science. The agreement has to be reached among people who are knowledgeable in the area concerned and who form their opinion by using the method of science. The agreed opinion must be arrived at after such individuals have either verified the result personally, or satisfied themselves adequately about the validity of the experiments and of the logic which led to the particular truth. The greater the above agreement and the greater the evidence in favour of a scientific truth (fact, theory or law), the more rigid must be the proof of evidence which will allow any modification of the truth to be accepted by the scientists, and the lesser will be the chance that the new evidence against the truth will stand a close scrutiny by them.

As time passes and the sum total of knowledge based on an initial consensus in the particular field increases, the chances that the initial consensus will prove to be wrong, diminish. This is the test of time in science.

For example, the chances that the proposition, "water consists of two parts (on molecular basis) of hydrogen and one of oxygen", will be proved to be wrong in the future must be considered as extremely remote, because a very large number of scientists—with probably no exception—have accepted this for centuries and have found no reason to alter their agreed opinion. On the other hand, there is yet no known general theory of cancer which may not easily be proved wrong in the future, because no theory has stood the test of time on the basis of an agreed opinion among experts.

Therefore, science is democratic, but not in a trivial sense. It does not allow one to change his opinion at the whims and fancies of a few.

An important objective of scientific training should be to enable one to evaluate evidence against an accepted and established scientific fact, theory or law, from the point of view of the probability of the evidence being true. It is this aspect of science which gives the method of science a built–in corrective: this corrective is provided through (a) the right to question, and (b) the requirement for a consensus.

There are no Know-alls in Science

Science does not claim to provide immediate answers to every legitimate question that can be asked at any given time. For example, we do not know yet what the mechanism of memory is (that is, how our brains store, collect and recall information), and what the ultimate fundamental particles are of which the entire matter in the universe is composed.

A scientist can say without any feeling of guilt or shame, "I do not know". What science considers undesirable is not the lack of knowledge, but the lack of desire to learn. And what goes entirely against the grain of science is the acceptance of answers that are absolutely incompatible with knowledge acquired through the method of science.

Science versus Supernatural

Science, therefore, denies the existence of the supernatural and of miracles. One often witnesses or hears about events which, in the opinion of many people, can have only a supernatural explanation, that is, an explanation outside the scope of the method of science. In reality, all such events do have a scientific explanation, often simple and ingenious!

For example, the famous magician, Houdini, could swim ashore safely after he had been tied in a sack placed in a trunk, the trunk locked and then dropped on a river bed. If he had started claiming supernatural powers and set himself up as a godman in our country, he would probably have acquired an enormous following! Here is an explanation of how he did it.

It is well-known that some people can develop, by training, special physical abilities, like the ability to twitch one's ear at will. Houdini had developed the physical ability of vomiting out the contents of his stomach at will. He swallowed a knife and the key to the trunk, just before he was tied up; physical search, therefore, never revealed him to be in possession of these objects (they would have, of course, been discovered on him—or rather, in him—if some one had decided to use X-rays!). To come out and swim ashore, he only had to vomit the gadgets out, cut open the sack and then open, from inside, the lock of the trunk.

If someone produces a rabbit out of a hat, or a watch out of 'nowhere', he is engaged in a trickery. All magic is trickery, with relatively simple scientific explanations.

Science does not seek an explanation of the unknown in terms of another unknown. When a scientist does not know the answer to a question, he says, "I do not know". He does not accept an "unscientific" answer, that is, an answer which cannot be tested or verified scientifically and which is incompatible with the method of science. He tries to find out if a scientific answer already exists. If no such answer exists, he uses the method of science to obtain an answer. His basic premise always is that, if the question is answerable, the answer can be found only through the method of science.

Predictability

A scientific fact, theory or law allows one to make testable predictions. For example, the Russian chemist, Mendeleev, predicted in 1869 the existence of several elements such as Gallium, Scandium and Germanium, long before they were discovered, and assigned to them their right places in his periodic table. Darwin and other evolutionary biologists predicted the existence of certain species, such as *Latimeria* (a fish), *Pithecanthropus erectus* (the "erect ape man") and *Oreopithecus* (another ancestor of man), much before their discovery. In physics, the existence of fundamental particles omega minus and the neutrino was predicted beforehand. In astronomy, the planets Neptune and Pluto would probably never have been discovered if astronomers hadn't looked for them following the prediction of their existence. And in space science, the first manned mission to the moon might very well have been a failure if scientists had not predicted with certainty that, on the moon, objects weigh one-sixth of what they did on earth!

It is possible today to predict solar and lunar eclipses to the fraction of a second.

We can predict that elements with atomic numbers 114 and 126, when discovered, will be found to be stable in contrast to all other transuranium elements—that is, elements with atomic number greater than 92 which have been discovered since 1940 and found to be (as predicted) unstable.

One can today predict with a fair degree of certainty the path of a cyclone and the existence of underground natural resources such as water, minerals, oil and coal. And one can predict with absolute certainly that if both the parents have blue eyes, all the children must have blue eyes. If a child does not, one of the parents must have committed adultery!

Scientific Observations are Verifiable and Repeatable

They do not depend on the whims and fancies of individuals. Thus, a time reaction set for 19+1 seconds, will take 19+1 seconds—no more and no less—irrespective of:

- who carries out the reaction (a child, a man or a woman—Indian, Chinese, African or European);
- what you may personally like to happen;
- the place where the experiment is done;
- the time of the day, month or year when the experiment is done.

[It would, of course, be necessary to ensure that the experimental conditions—such as temperature—which are known to influence chemical reactions in accordance with scientific laws, are maintained constant.]

Science is Truly International

There is one and only one science. Scientists all over the world use the same method (the method of science), employ the same techniques, use the same materials, publish in the same journals, are increasingly beginning to use the same language (that is, English), and form a truly international community in which the professional links are at least as strong, if not stronger, than other links.

The method of estimation of protein described by Lowry in 1951 has been—and continues to be—used widely by scientists all over the world. In 1972 alone, at least 5925 scientists from every country where any significant scientific research is done, employed this method.

In 1951, scientists took out some cells from a cancerous tumour in a woman called Henrietta Lacks. The cells were then grown outside of her body in the laboratory. She died of cancer but her cells continued to multiply in the laboratory—first in one laboratory, then in a few more laboratories, and so on. Today, these cells (HeLa cells as they are called) have found their way in research laboratories all over the world. Henrietta Lacks, in a way, continues to live, and not just at one place but at many! And scientists all over the world today use the same strains of mice, rats and bacteria so that they can compare their results.

Every country in the world today where science is done—except possibly Soviet Union—publishes journals which have papers in English, because scientists want to understand each other.

Compare science, which is truly united at the international level, with other activities (such as astrology) that are widely practised and believed in and do not possess such a unity. If you were to compare astrological predictions for the same week and forecast for the same set of birthdays made by different people in different countries, you would notice how much these forecasts differ from one another.

Differences arise in science too, but the scientists make a genuine effort to resolve them.

Science and Religion

This discussion would be incomplete without comparing—or contrasting—scientific and religious beliefs. A comparison of science with religion is necessary for the following reasons. (The term religion here refers to religious dogmas that are responsible for the identity of a religion, and not the value system of religions which is essentially the same for all religions and is in agreement with science.)

(a) Religion has attempted to answer the same questions as science has.

(b) Explanations provided by religion are usually older than those provided by science.

(c) Science became in history a competitor to religion and often came into direct conflict with religion. However, whenever such a conflict arose, the explanation provided by science through the use of the method of science was always found to be correct and was eventually accepted.

(d) Religion has often hindered the progress of science, that is, of knowledge, and persecuted scientists.

All new knowledge in science must be consistent with known and established observations. Science progresses through modification of a part of the existing knowledge and not by replacement of the entire body of knowledge. Growth of scientific knowledge is a continuous process; science is, therefore, evolutionary.

Contrast the above attributes of science with the following attributes of religion:

(a) Religious dogmas, including the so-called miracles (for example, the materialization of objects by a wave of hand), are often inconsistent with known and established observations.

(b) A new religion attempts to replace fully the existing religions.

(c) A religion once founded continues substantially unchanged.

Religious texts, on which the followers of religion depend, are generally ancient. The founders of the religion who lived in the remote past, matter most in religion at any time. For the followers of religion, religious events of the past are the events of the greatest concern. Religious customs and practices do not basically change with time. Whatever changes are brought about are due to forces external to religion, such as science itself.

For the followers of science, the more modern the text is, the better it is. The scientists of the time matter most in science at a given time. For the followers of science, the events of today and the likely events of tomorrow are the events of the greatest concern. The techniques used in science keep on improving with time, and the impetus for improvement comes from within the framework of the method of science.

What is inspiring in science is not what had happened in the past, but what is happening now or what may happen in the future. In religion, by contrast, inspiration comes almost entirely from events of the past.

There are many religions and they differ from one another in many respects. Activities of a religion are carried out in isolation of other religions, people from other religion being often prohibited from participating in it. There is little or no communication among various religions and, therefore, no common language. Religious practices differ enormously—often fundamentally—from religion to religion; they divide people. Science, on the other hand, is unitary, open and truly international, and helps unite people.

Science, unlike religion, takes one forward.

Our Obligations

In the end, let us look at what our obligations are, if we accept the method of science as something that works. The acceptance of the method of science implies, (a) an understanding of the basic attributes of the method; (b) acceptance of knowledge gained by its application as the closest available approximation to truth; (c) rejection, at a given time in history, of what is incompatible with the above knowledge; (d) recognition that the method of science is the only way of acquiring knowledge; and (e) application of the method of science in everyday life and in every sphere of human activity. Not an easy recipe, but perhaps the most satisfactory one available today.

XI

DOES SCIENCE REFUTE RELIGION?*

P M Bhargava

Science and religion both concern us, and intimately so. For, in the world of today, we cannot do without science, and religion has been woven into the very fabric of our society for centuries. But, then, you may ask, what *is* the problem? Indeed, there would be no problem if they—religion and science—complemented or supplemented each other. It turns out that they do not, even though many would like them to; some people even make valiant attempts to make it appear so, regardless of the truth.

What I would like to do in this write-up is to dwell upon this contradiction—that is, the contradiction between science and religion. What I hope to be able to show is that there is an *inherent* incompatibility between science and certain aspects of religion, especially the dogmatic part that gives a particular religion its identity.

Evolution of Religion and Science

One may, of course, ask: why is it necessary to compare or contrast religion and science? It is necessary to do so for at least one important reason, that is, the fact that religion, all through the history of man, has attempted to answer the same questions as science has. The origins of both religion

* This article appeared in *Society and Science*, 1981, Vol.4, pp.42-50.

and science can be traced to the evolution of intelligence in man. Intelligence is just another name for the ability to ask questions. One can, therefore, surmise that, in the remote past, when man came to be endowed with intelligence, he must have asked himself questions—questions of at least four kinds. First, questions about the non-living materials he saw around him, such as water, air, earth and minerals. Secondly, questions about the physical phenomena he witnessed, such as light, heat, sound, thunder and lightning. Thirdly, questions about the extraterrestrial objects and phenomena he observed, such as the periodical rising of the sun, the moon and the stars; the passage of the planets through the various constellations and, of course, the eclipses. And fourthly, questions about the living things that he saw around him, for example, the recurrent phenomena of birth, disease and death.

All these phenomena, which now come under the purview of the four basic sciences—chemistry, physics, astronomy and biology—must, indeed, have intrigued early man. How did he then go about finding the answers? The method of science had not developed, and the whole logic and logistics that we have today for answering such questions, did not exist at that time. What did the primitive man then do? He used his intelligence to construct self consistent systems of beliefs such that once you accepted certain premises entirely on faith, and without questioning, answers that were plausible, at least at that time, emerged. It is this kind of effort that perhaps led to the development of religion, both pagan and codified—the codified religions including Hinduism, Buddhism, Judaism, Christianity and Islam. However, as the total fund of human knowledge increased, a time came when man began to *question* the basic premises of religion. Out of this questioning, perhaps, crystallized what we today know, formally, as the method of science. It soon became apparent that this method could not only be used as a tool which would satisfy human curiosity much more than religion had done so far, but it also opened up new areas for investigation that had so far been hidden or even prohibited. The phenomenon snowballed from the thirteenth century onwards, and we had Roger Bacon, Leonardo da Vinci, Copernicus, Francis Bacon, Galileo, Rene Descartes and Newton, amongst others, to give new dimensions to the method of science—to the newly developed art of questioning. The answers that emerged did not demand acceptance on the basis of faith alone; moreover, they were testable and verifiable, and did not depend on the whims and fancies—or the likes and dislikes—of an individual or a group of individuals. Soon, science became a competitor to religion and, often, came into direct conflict with it. We had in the 16th and 17th centuries, the conflict between Copernicus and Galileo on one side and the Church on the other. More recently, in the last century, the Church waged another major battle—this time against the Darwinian theory of evolution. which was so ably extended by Thomas Huxley to the evolution of man.

As science opened up new vistas, religion soon became a hindrance to its progress, and led to the persecution of scientists. Copernicus had to recant because he said that it was not the sun that goes round the earth, but the earth that goes round the sun. Galileo, a follower of Copernicus, died under house arrest on account of holding on to Copernican beliefs. And, before Galileo, Bruno was burnt at the stake for reasoned dissent. As recently as a hundred years ago, Darwin and Huxley were laughed at by an uneasy Church for saying that man has evolved from 'lower' creatures, and not put on the earth as an act of creation. However there was a redeeming feature for science too. Whenever a conflict arose between science and religion, the explanation provided by science through the use of the method of science, was eventually *always* found to be more appealing to reason.

Science, therefore, grew up, so to say, as a competitor to religion, and the battle still continues on many fronts. Clearly, then, there is sufficient justification for comparing (or contrasting) science and religion.

Let us now get on to specific points of comparison and contrast. A good way to do so would be through a comparison of the definitions and the attributes that are widely accepted, of both religion and science—hoping, of course that semantics will neither rule the roost nor come in the way.

Attributes of Religion and Science

Religion is defined variously as service in adulation of God as expressed in forms of worship, a system of faith and worship, and an awareness or conviction of the existence of a supreme being that arouses reverence and the will to obey. The existence of the supernatural—that is, something which is beyond the laws of science—is implicit in religion, no matter what definition one accepts. In all religions, there is also provision for the supernatural to take the form of what appears to be natural. Thus, Messiahs or Avatars are born on this earth, and god takes the shape of man or even other creatures, as is supposed to be the case with some of the incarnations of Vishnu. It is this inherent belief that underlies religion, which has led to the emergence of various forms and shapes of godmen—be it Maharishi Mahesh Yogi, Shri Satya Sai Baba, Acharya Rajneesh, or what have you. These godmen would like others to believe that they have supernatural powers which cannot be understood by other men, and that their statements and actions must, therefore, be accepted by others without questioning. Science, on the other hand, does not accept the existence of a high priest, a godman or any other authority that cannot be questioned. In fact, science denies the existence of the supernatural and of miracles, which are the very essence of religion. One often witnesses or hears about events which, in the opinion of those who are religious, can have only a supernatural explanation—that is, an explanation outside the scope of the scientific method. In the view of science, all such events—assuming they have ever occurred (which, at times, is doubtful)—do have a scientific explanation, often simple and ingenious.

Religion is based on revelation. Indeed, revelation is *the* method of religion. Truth was revealed to, and not discovered by, all the religious leaders of the past—be it Moses or Mohammad, Christ or Buddha, Mahavir or Aurobindo. The method of science that the scientists use, has no place for revelations of that kind. It consists of distinct steps: the framing of a question on the basis of careful observation or a careful analysis of existing data; the formulation of a *testable* hypothesis; the doing of experiments; and, finally, arriving at the answer by using existing knowledge and logical reasoning. One may, of course, in certain circumstances, omit one or the other of the steps of hypothesis and experiment, but that is not really important. What *is* important is that the method of science has a built-in corrective and that the conclusions that one arrives at by using this method are verifiable and repeatable. For example, I can set up a time reaction in which two colourless solutions when mixed with each other at time zero, will turn violet all of a sudden at 24 seconds. It will take exactly the same time (within experimental error) for the mixture to turn violet, no matter who does the mixing—a child, a man or a woman; an Indian, a Chinese or an African; a Brahmin or an untouchable; a Punjabi or a Tamilian—and no matter what you may personally wish to happen. Neither a particular confluence of planets, nor the will of all the godmen in the world, can increase this timing of 24 seconds to, say, 44 seconds, or decrease it to, say, 4 seconds.

Another important attribute of science is that it allows one to make testable predictions. It was the scientists' ability to make predictions with considerable certainty and accuracy, that allowed man to land on the moon. Indeed, in the one grand experiment that the first manned landing on the moon constituted, an enormous number of predictions made by scientists in a vast variety of fields came true. If one of these predictions had gone wrong, there would have been disaster. Science has, in fact, allowed us to predict chemical elements, fundamental particles, planets, biological species, and a host of other exciting new objects and phenomena. For example, the great Russian chemist, Mendeleev, predicted, in the middle of the last century, the existence of the elements gallium, scandium, and germanium. For gallium, he said that when the element is discovered, it would be found to melt with the heat of the hand. Its melting point, when the element was actually discovered, was found to be about 30°C. Pauli, in 1930, predicted the existence of the fundamental particle, neutrino, which is now very much in the news. It was discovered in 1956. Murray Gellmann who, like Pauli, was awarded a Nobel Prize, predicted the existence of another exotic fundamental particle, omega minus, in 1962. This particle was discovered later by Samios.

The existence of planets, Neptune and Pluto, was predicted. In fact, in the case of Pluto, the exact place in the sky where it should be found, on the fateful night in March 1930 when it was discovered, had been correctly predicted. Darwin's theory of evolution predicted the existence of *Latimeria*, a fish, and *Pithecanthropus erectus* and *Oreopithecus*, two ancestors of man. Both the fish and the human ancestors were discovered subsequently—the fish as a live specimen and our ancestors as skeletons. Both were found to have the characteristics predicted by Darwin and his followers. Today, we can predict that elements with atomic numbers 114 and

126 when discovered will be found to be stable in contrast to all other transuranium elements, that is, elements with atomic numbers greater than 92, which have been discovered since 1940 and found to be, as predicted, unstable. In the entire history of religion there has not been a single such prediction, made on the basis of religion, that has subsequently come true. Here is, therefore, an important point of contrast between science and religion: science can allow us to make testable predictions; religion cannot.

Dynamism of Science

In science all truths are truths by consensus. Of course, the consensus has to be reached among people who are knowledgeable in the area concerned and have formed their opinion by using the method of science. The consensus must be arrived at after such individuals have verified the results personally, or satisfied themselves adequately about the validity of the experiments and of the logic which led to the particular truth. As time passes and the sum total of knowledge in the particular field increases, the chances that the initial agreement of opinion will prove to be wrong, of course, diminish. Nevertheless, I repeat, all truths at any given time in science are truths by consensus—a consensus among scientists. On the other hand, religious truths represent an opinion usually of one religious leader, at most of a few. Moreover, these opinions are rigid. Changing them implies establishing another religion, or at least a sect. Therefore, a given religion, by definition, is static, unlike science which is dynamic and changes with time as more and more evidence comes forth. In fact, it is often said that science progresses by disproving. At least two Nobel Prizes were awarded for discoveries which were subsequently proven to be incorrect. However, in both these cases the persons concerned deserved to receive the Nobel Prize because, had they not made their discovery, the truth as we know it today, would not have been discovered at the time it was. Science is, therefore, evolutionary, which religion is not. Science has a built-in corrective, which religion does not have.

In science, a new theory must explain all that was explained by the old theory plus something that could not be explained by the earlier theory. The new theory, in addition, should be capable of making predictions which could not be made on the basis of the old theory, and some of these predictions should, indeed, have been tested and turned out to be right. For example, Einsteinian physics made predictions which Newtonian physics could not, and explained events and phenomena which the earlier physics could not. Thus, the interconversion of mass and energy, the bending of light in the presence of a large gravitational field, the existence of black holes, and the dependence of the mass of an object on its speed, were all predicted by Einstein and later on substantiated. None of these predictions was possible on the basis of Newtonian physics. That is why we consider Einsteinian physics an improvement over Newtonian physics from which it actually evolved. Contrast this situation with what is obtained in religion, where no religion can be said to be an improvement over any earlier religion. If you say something to the contrary—that one religion is an improvement over another—you might have a riot!

All new knowledge in science must be consistent with known and established observations. On the other hand, religious dogma (including the so called miracles, for example, the materialization of objects by the wave of one's hands) is often inconsistent with known and established observations.

Science progresses through modification of a part of the existing knowledge and not by replacement of the entire body of existing knowledge. A new religion, on the contrary, often attempts to replace fully the existing religions.

The growth of scientific knowledge is a continuous process. Science is, therefore, evolutionary. A religion once founded continues substantially unchanged.

Religion and Science: Some More Differences

Another important difference between science and religion is that while science is forward-looking, religion is backward-looking. For example, for the followers of science, the more modern the text, the better it is. On the other hand, religious texts on which the followers of religion depend, are, generally ancient. In the case of science, the scientists of the present time matter the most; in the case of religion, the *founders* of the religion who lived in the remote past matter the most. For the followers of science, the events of *today* and the likely events of *tomorrow* are the events of the greatest concern; for the followers of religion, the religious events of the past are the events of the greatest concern. The techniques used in science keep on improving with time, and the impetus for this improvement comes from within the framework of the method of science. On the other hand, religious customs and practices do not basically change with time. Whatever changes are brought about are due to forces external to the religion—such as science itself.

An important attribute of science is the right to question. Knowledge advances and science progresses *because* people exercise their right to question. By contrast, religion demands an unquestioned acceptance of its tenets and dogma. If you question, it must be only to seek clarification and not to doubt.

A scientist can say without any feeling of guilt or shame, 'I do not know'. It would be disastrous for a religious leader to say 'I do not know'; he would simply lose his following. By definition he knows all. Every major religious leader of the past—the founder of every religion—had answers to every question that one may ever ask. Science would consider such a claim as hypocrisy and deceit.

Internationalism and Parochialism

Another important difference between science and religion is that while science is truly international, religion is not. Scientists all over the world use the same method, that is, the method of science. They employ the same techniques, use the same materials, and publish frequently in the same journals. They are increasingly beginning to use the same language—that is, English—and they form a truly international community in which the professional links are at least as strong as any other link. Contrast this internationalism of science with the parochialism of religion. There are many religions and they differ from one another in many respects. The activities of a particular religion are carried out in isolation of the other religions: in fact, people of other religions are often prohibited from participating. There is little communication between various religions and, therefore, no common language. Religious customs and practices differ enormously, often fundamentally, from religion to religion. Religion, in fact, divides people while science unites them.

Having said all this, one may now ask 'what about values', which, indeed, are an integral part of all religions. Aren't they good?—that is, good for us. Indeed they are; but such values are not a special characteristic of any particular religion. They are, in fact, common to all religions. Every religion asks you not to kill your fellow men, to be kind to them, to care for them. Every religion prescribes human compassion, truthfulness, integrity and honesty. However, a particular religion receives its identity not from these values, but from its dogma. A religion bereft of its dogma is no longer a religion. It is, in fact, the dogmatic part of religion that contradicts science.

I must, here, of course, mention that dogma does not arise from religion alone. Custom, convention, tradition—occasionally science too—may lead to the establishment of a dogma. It is just that religion has been the most important source of dogma. Therefore, one is justified in saying that religion and science do not mix.

I must here also add that most values derived from religion that are universally cherished, are compatible with science. These values can be arrived at through a scientific argument as well. In fact, one advantage of using science and not religion as the basis of determining whether a value system is beneficial to man or not, is that in science nothing is immutable. Therefore, the scientific assay for a value system would allow a change in it—a change which emerges logically and naturally from the environment. Religion, by contrast, demands immutability of a value system.

Conflicting Views

Towards the end, let me cite some specific examples of contradiction between science and religion. Today, we understand, reasonably well, what might have been the likely origin of the universe. It is generally accepted by scientists that the universe came into existence about 14 billion years ago, and they can trace the history of the universe backwards to nearly 10^{-42} of a second just after the event of its formation, called the 'Big Bang' by astronomers. In this scheme, there is no need to postulate the existence of God, as one must do in religion, unless we ascribe to God just one function: that is creation of the Big Bang. He would then have been rendered redundant after the first 10^{-42} seconds following the Big Bang as, later, the laws of science took over, over which he had no control!

Today, scientists can say with considerable certainty, that life on our planet evolved from non-living materials. After the formation of our Earth, complex chemical substances were slowly formed from the simple chemical substances that were contained in the primordial atmosphere; such a 'chemical' evolution eventually dovetailed into the biological evolution leading to the appearance of our species on Earth. On the other hand, all religions demand the acceptance of the belief that man (and, in the case of some religions, other forms of life as well, including women) were put on Earth by God through a deliberate act of creation.

Some of the religious leaders of the past were supposed to have been born through immaculate conception—an idea which is utterly incompatible with scientific truth about reproduction. And virtually every religion postulates some kind of life after death. The concept of soul is common to all religions. On the other hand, a scientist may ask the question: "Where has the soul been if you can bring a dead man back to life, as indeed you do when you take out the heart of a person and replace it with the heart of another person?" Soul and rebirth have no place in science.

Scientific Laws

Today, it is possible to grow a whole plant from a single cell of the plant. Such cloning is, no doubt, theoretically possible even for higher organisms, such as man, and it is only a matter of time before it will be done. It is now also possible to fertilize a human egg with a human sperm in a test-tube, under controlled conditions. We understand a great deal of what happens during such processes. In fact, today, one can say with confidence that all phenomena associated with life and the life processes, must have an explanation in the laws of physics and chemistry. Indeed, if one can explain satisfactorily the life-associated phenomena, the extra-terrestrial phenomena, the physical phenomena, and the nature of the non-living materials, and if one can provide a scientific basis for values, where is the need for religion?

On the contrary, history tells us that unquestioned acceptance of religion, and of all that religion demands, leads to perpetration of untruths. In contemporary history, it is religion that has wooed science and not vice versa. Maharishi Mahesh Yogi and so many others of his breed, all the time seek the blessings of science in support of their claims. Scientists never need to invoke religion in support of *their* claims. Scientific truths do not need a prop. Religion does—at least today!

To end, let me say that one important fundamental characteristic of science is that it is not evangelistic; it does not seek to convert people, unlike many religions. People must, therefore, be permitted their beliefs, religious or otherwise, in any free society, and I will stake everything to defend the right of every individual to believe in the religion he wishes—as long as his beliefs do not hurt others. However, to say that religion and science are compatible or that one derives from the other or supports the other, must be considered wishful thinking at best, and a travesty of truth at worst.

XII

SCIENCE AND DOGMA*

P M Bhargava

In contemporary life, we have to constantly face contradictions between science on the one hand and dogma of various kinds on the other. Dogmatic beliefs, like science, have come to be a part of our very existence. Arising out of unquestioned acceptance of religion, custom, convention or tradition, they have, over generations, acquired an authority out of all proportion. Fortunately, of course, everything in religion, custom, convention and tradition is not dogma.

A well-known example of the contradiction between science and dogma, relates to our own origin that is, the origin of man. Modern biology tells us that life has evolved on this earth from non-living things as a result of evolution of complex chemicals from simpler ones: an evolution that slowly developed into the biological evolution. On the other hand, all religions tell us that man has been put on this earth as preformed man, an act of creation by the will of God. The two opinions are entirely and absolutely contradictory. Which one should one believe in? If in your mind dogma takes precedence over science, you will probably classify yourself as a creationist, if on the other hand you have a scientific approach you are likely to be an evolutionist. Let us see why.

Dogma, by definition, is an opinion or a tenet accepted

---

* This article appeared in *Link*, August 15, 1982, pp.171-173.

entirely on the basis of faith, without questioning. Questioning is simply not allowed by the adherents of dogma. They may seek clarification, but never question a dogmatic belief. The contradiction with science begins right here. Science does not accept anything without questioning and entirely on the basis of faith. The method of science is totally incompatible with the view that a man or an authority exists—or has ever existed—whose opinion must be accepted entirely on faith, without questioning. In other words, science does not accept the existence of high priests of any hue or colour. This is, of course, not to say that there are no self-styled high priests in science. It is only to make the point that whether or not they like it, in science they are bound to be questioned—and often by people of seemingly lower status, for there is no tenable hierarchy of status in science.

Much of the dogma we see around us arises out of religion. The problems become acute because the origins of religion and science are about the same. Probably both arose as a result of the inbuilt desire in man to find answers to question for example, questions pertaining to natural phenomena. What are the nonliving materials that we see or feel around us all the time, such as air, water, minerals and rock, made of? What is the nature of the physical phenomena that we witness so frequently, such as heat, light, sound, thunder and lightning? What is the basis of extra-terrestrial phenomena that have been known for so long, such as the periodic rising of the sun, the moon and the stars, and the occurrence of eclipses? And what is the nature of the phenomena we see associated with life, like birth, disease, growth and death? Questions such as these must have been asked by primitive men.

Probably, it is the desire to find answers to questions like these that provided one of the important motivations for the formation of religions—pagan or codified (by codified religions, I mean major religions such as Judaism, Hinduism, Islam, Buddhism and Christianity). Since the scientific method that took formal shape only some eight centuries ago, has also attempted to provide answers to the same questions, it is clear that both religion and science have treaded the same ground; often, the two have been competitors. On account of differences in the fundamental attributes of religion and science, the nature and character of the answers provided to questions by the two approaches have been different—often contradictory. In the case of religion one sub-set of these answers is represented by dogmatic beliefs. Hence the conflict between science and dogma.

Let us continue our comparison—rather contrasting—further. Scientific answers are verifiable, repeatable, and objective, and do not depend on the whim, fancy or desire of a particular individual. This is not true of answers that have come to us as dogmatic beliefs. Such beliefs represent the fear of the unknown. Whereas scientific answers represent an understanding of the known. In the case of dogmatic beliefs, the last word has already been said. In the case of answers arrived at through the application of the method of science, the last word shall never be said as all scientific answers are tentative and the results of a consensus amongst the scientists of the time. The consensus is, of course, arrived at through the application of the method of science.

All new knowledge in science must be consistent with known and established observations. Dogmatic beliefs, including the so-called miracles (for example, the materialisation of objects by a wave of hand) are often inconsistent with known and established observations. Science progresses through modification of a part of the existing knowledge and not by replacement of the entire body of knowledge. On the other band, a new set of dogmatic beliefs often attempts to replace fully the existing sets of such beliefs. There is thus no overlap between Christian and Hindu dogma. Growth of scientific knowledge is a continuous process and science is, therefore, evolutionary; by contrast, a set of dogmatic beliefs once founded, continue substantially unchanged.

Texts on which the followers of dogmatic beliefs depend, are generally ancient. For the followers of science, the more modern the text is, the better it is. For those who believe in dogma, the founders of the dogma who lived in the remote past are the events of the greatest concern. For the followers of science, on the other hand, the events of today and the likely events of tomorrow are the events of the greatest concern. Customs and practices arising out of dogmatic beliefs do not basically change with time, whatever changes are brought out, are due to forces external to these beliefs, such as science itself. By contrast, the techniques used in science keep on improving with time, and the impetus for the improvement comes from within the framework of the method of science. In other words, while dogmatic beliefs are backward-looking, science is forward-looking.

After this exercise in comparing science with dogma, let us look at some more examples taken from our everyday life where we are required to make a choice between the dictates of science and those of dogma. Belief in rebirth and soul is a widely accepted dogma even though there is no reliable evidence either in favour of rebirth or in favour of the existence of soul. In fact, what we know of modern biology makes rebirth as we understand this term in common parlance, not only untenable but absolutely impossible. The same can be said of the concept of soul. It is, therefore, understandable why probably over 99 percent of the distinguished biologists of today around the world categorically reject the existence of either the phenomenon of rebirth or soul. In fact, it is belief in rebirth and soul that gives credibility to a host of social evils such as the caste system, in the minds of people. If you are born in a lower caste, only you are to blame. It is your deeds in the past life for which you are now paying the penalty—something which no one can prove!

A sense of superiority on account of circumstances of birth that is reflected in communalism, casteism, stateism or linguistic chauvinism, is also a result of the dogmatic belief that the circumstances of birth, rather than the genetic make-up of an individual are the primary determinants of a person's ability, capability or competence. Such beliefs go against the very grain of modern biology which, during the last 30 years, has allowed us to understand the basis of similarities or dissimilarities in the entire living universe. We today understand the mechanism of heredity and the nature, structure and function of the genetic material, that is, DNA. Fatalism also arises out of such dogmatic beliefs. It deters people from exercising their rights and from raising their voice against inhumanism, oppression and exploitation.

Let us take another example: that of the concept of death. According to all religious beliefs, death is a unitary event. When a person dies, he simply dies irrevocably. He is, therefore, either alive or dead. Modern biology has shown us that this concept is inaccurate. Even though a person may die, for example as a result of his heart becoming non-functional, he can be brought back to life by putting in another heart. An organism may die but certain tissues of it may stay alive. Even if a tissue dies, its cells may not die and could be taken out and maintained for a very long time (if not for ever) in the laboratory. If one remembers that each cell of a living organism carries the entire blue-print for the whole organism, and if we can maintain the cells of a person who is otherwise dead and gone, we theoretically retain the possibility of creating an identical person from any of these cells. It has indeed been possible to do so in the case of plants. One can today raise a whole carrot plant from a single cell of virtually any part of the carrot plant.

The everyday life of many of us, indeed, centres around a host of dogmatic beliefs. For example, a large number of people in our country believe in the existence of heaven and hell, and in astrology. These beliefs satisfy all the criteria of dogmatic beliefs that stand out in straight contradiction to the sum total of scientific knowledge we have today, Dogmatic beliefs are widely prevalent in regard to what we may eat and how should we go about curing a particular disease, which beliefs, again, contradict the knowledge we have acquired about nutrition and about health and disease through the application of the method of science. We cling to these beliefs tenaciously and work up fantastic arguments to show that they are compatible with science, because science is something we cannot reject.

Let me end by saying that the pressures on us to cling to these beliefs that prevent us from progressing, come from the privileged because it is in their interest that people continue to hold these beliefs and thus remain ignorant. Indeed, if the masses were to shed their ignorance through proper education that encourages them to question such beliefs, and retain only what appeals to their reason, those who are privileged on account of circumstances of their birth would soon cease to have their privileges.

Our salvation lies not in continuing to adhere to dogma, but in following the precept of Lord Buddha who said: "Believe nothing merely because you have been told it or, because it is traditional or because you yourself have imagined it. Do not believe what your teacher tells you merely out of respect for the teacher, but whatever after due examination and analysis you find to be conducive to the good, the benefit, the welfare of all beings, take it as your guide."

XIII

HISTORY OF THE EXHIBITION ON THE METHOD OF SCIENCE*

P M Bhargava

I. THE BEGINNING

It all started with Rais Ahmed, a physicist from Aligarh Muslim University, who became probably the most effective Director so far of the National Council of Educational Research & Training (NCERT). A rationalist and a nationalist to the core, he was responsible for the 10+2+3 formula for our education, which has since then been introduced all over the country; he was also responsible for formulating our national policy in respect of vocationalisation of education through two streams after high school, one that would go towards vocational education and the other that will go towards university education.

We have a national science exhibition in Delhi every year on Jawaharlal Nehru's birthday, the 14th November. When I was in Delhi on 14th May 1975, I went to see Rais Ahmed, who was then quarantined at a place at the airport

* This article is Part 1 of the book *Vandalisation of a Work of Art and Science: A Documented History of P M Bhargava's Famous Exhibition on The Method of Science, 1975-2002*, edited and published by B Premanand, 2005, for Geedee Medical Aids.

for entering India from Africa where he had gone on a short visit, without having had yellow fever inoculation. He then said that as I had been talking and writing about scientific temper and the scientific method for the last two decades, I should consider setting up an exhibition on the method of science as a part of the national science exhibition. He said that if I agreed to do that, the NCERT will support it financially.

The idea appealed to me as it was intellectually challenging. To present in a visual form something as abstract and esoteric as the scientific method, in a way that it would appeal to both young and old, required a high level of creativity. I, therefore, accepted Rais Ahmed's invitation.

As soon as I came back to Hyderabad, I stayed up one night and, in one go, wrote the first draft of the entire 50-page script of the proposed exhibition. During the next few days, I worked out roughly what may be needed in terms of audio-visual material and live experiments to convey what was written in the text; the idea was that the text must not dominate the exhibition. All of the above—the text and the audiovisual material (including live experiments and demonstrations)—went through several revisions but the basic format conceived in the first few days stood the test of time and stayed unchanged!

I then contacted the following persons in Hyderabad who were acknowledged experts in their own field, to seek their cooperation in putting the exhibition together by contributing specially in the area mentioned against each:

Text	Manorama Bhargava
Drawings & Paintings	Laxma Goud
	A Shareef
Films & Video	P V Satheesh
Photography	M Hasan
	Kaiser Jamil

Experiments & Sound Effects	M W Pandit
Models & Instrumentation	P S Kolhatkar
	B Gopinath
	Sanjar Ali Khan
Construction & Display	Surya Prakash

To my great delight, all the above agreed, and the first meeting of the above group was held at my house in the campus of then Regional Research Laboratory (RRLH), now the Indian Institute of Chemical Technology (IICT), from 9 PM to midnight on 28th June 1975. I believe the participation of the above individuals, some of whom are very well-known today in their respective areas of activity, in putting together what is now known as the Method of Science Exhibition (MOSE), played the same role in transforming their own lives as it did in my own.

I realised that the money that the NCERT was to offer wasn't going to be enough for the Exhibition. We also concluded that for the Exhibition to be meaningful, it would need to be far bigger than what may be suitable as a part of the national science exhibition. I, therefore, took two decisions. One was to look for money outside of the NCERT with, of course, the approval of the NCERT. The second was to persuade Rais Ahmed to think of putting up the Exhibition on a permanent basis as a separate entity and not as a part of the temporary yearly national science exhibition. Rais Ahmed agreed to this idea.

As regards raising funds, I approached, in August 1975, Anand Lal, the then Chairman of VST Industries. I was delighted that the VST decided to make, in 1975, a grant of Rs. 1 lakh towards the cost of the Exhibition. Had the VST not provided this funding, the Exhibition would have never come about. I also received total support from the RRLH where I was working from 1975 to 1977. On 1st April 1977, the Centre for Cellular and Molecular Biology (CCMB) was created as an independent constituent of the RRLH; it subsequently became a separate constituent laboratory of the CSIR. The CCMB also provided unstinting support to this Exhibition as did our parent body, the CSIR which was then under the leadership of Y Nayudamma.

As it turned out, there was a building available in the IICT campus which had been built for conducting courses for the University of Hyderabad before this university had its own buildings. These courses had finished and we could make use of this building to prepare and keep our exhibits in the campus of the RRLH.

We received enormous support from numerous individuals and organisations around the world who provided us valuable and rare display material. Some of these were the following:

Review of the text
Prof. Shiv Kumar, University of Hyderabad
Dr. Amritjeet Singh, American Studies Research Centre, Hyderabad

Testing of the text
Shri Kumar Ketkar, Bombay
Many friends, young and not-so-young, from Hyderabad, who volunteered to have the text tested on them.

Printing of the text
Impress, Bombay

Spoken words in the films, the videotape and the limericks
Shri K P Mohanan and other members of the staff, and students, of the Central Institute of English and Foreign Languages, Hyderabad

Photography
State Photo Service, Secunderabad

Information and material

Dr. Indradev, Administrative Staff College of India, Hyderabad

Kum. Safia Mehdi, Hyderabad

Smt. Vanaja Iyengar, Women's College, Hyderabad

Shri Henry David, Loco Shed, Secunderabad

Shri A R Shahjehan, Hyderabad

Shri R Raghuveer Bhopale, Poona

Prof. S M Razaullah Ansari, Aligarh Muslim University, Aligarh

Shri A Rahman, CSIR, New Delhi

Dr. G S Aurora, North-Eastern Hill University, Shillong

Prof. E A Dawes, University of Hull, Hull

Jagdish & Kamla Mittal Museum of Indian Art, Hyderabad

National Geophysical Research Institute, Hyderabad

Max Mueller Bhavan, Hyderabad

Andhra Pradesh Agricultural University, Hyderabad

M N J Cancer Institute of Radium Hospital, Hyderabad

South Central Railway, Secunderabad

India Meteorological Department, New Delhi

Bhabha Atomic Research Centre, Trombay

Hindustan Antibiotics Limited, Pimpri

Virus Research Centre, Poona

The British Council, Madras

United States Information Service, Madras and Bombay

The Royal Society, London

Institut Pasteur, Paris

Many members of the staff of Regional Research Laboratory

Upagraha Doordarshan Kendra, and Hyderabad Science Society, Hyderabad

Many other friends and organisations from far and near

Many publishers and authors from whose books and papers some of the photographs and other non-textual material were reproduced.

It took us nearly two years to put together the Exhibition. As it was nearing completion towards the end of 1976, we realised that to display it effectively, the minimum space required would be about 5,000 sq. ft. plus some space for management. While it was being set up in Hyderabad, many people came to look at it. It was generally felt also by them that it should be set up in Delhi as a permanent Exhibition.

As both Rais Ahmed and the then Minister for Education, Nurul Hasan (whom I had the pleasure of knowing well for several decades earlier), were convinced that the Exhibition would be a major addition to the Indian intellectual scenario, they found a most appropriate place for the Exhibition in Delhi: the Polish Pavilion in the campus of the Bal Bhavan on Kotla Marg. This pavilion was a separate building located on about half-an-acre of land in the Bal Bhavan campus. The location was ideal, and it would allow the Exhibition to maintain its own identity.

I must add here that during this period (that is, from 1975 to 1977), while the Exhibition was being set up, we had some problem with Athindra Bose who headed the science division of the NCERT. I had known Bose for a very long time as he and I were class-fellows in BSc and MSc at Lucknow University from 1942 to 1946. I came in intimate contact with him in the NCERT also because, during the same period that the exhibition was being put together, I chaired a textbook-writing committee for the NCERT which led to the well-known text book, *Learning Science–Part I*, for class VI (this was the prescribed national textbook for class VI for science for many years and was the first text book in which an integrated approach was taken towards teaching of science). As the book had deviated from the norms in many ways, we had some problems with Bose in this regard as well which we, of course, eventually overcame. Bose had extreme rightist leanings which were well-known and which were totally contrary to the values of all of us who were concerned with the Exhibition.

The whole team that had put up the Exhibition came to Delhi to set it up at the Bal Bhavan. In addition, my wife and two children and another colleague from the RRL came along.

By the end of 1976, the Polish Pavilion area in the campus of Bal Bhavan in New Delhi was vacated for us and all the work we wanted to be done to have it readied for the Exhibition, had been done. We shifted all the material of the Exhibition to Delhi between 6th and 17th February 1977. We had to take a whole wagon for the nearly 2000 cartons of the Exhibition material, some of which we carried ourselves in the train. For this we incurred an expenditure which was a little more than what had so far been budgeted, and there were also payments to be made to all those who had worked for the Exhibition. I was, however, told by the NCERT that all these payments, including the payments due for travel etc., would be made by the NCERT later. We felt confident on this score when we were invited by the Education Minister, Nurul Hasan, to have tea with him in Shastri Bhavan on 7th March 1977; he patted us on the back and said that in his opinion the Exhibition was going to be a landmark in the intellectual history of the country. Our team was thrilled and worked with increased vigour to bring the Exhibition to a stage that it could be opened. The idea was that it would be opened by the then Prime Minister, Indira Gandhi. She had been made aware of the Exhibition by several people, including Nurul Hasan and Rais Ahmed. Further, while the Exhibition was being set up in Delhi, many dignitaries, such as the late Y Nayudamma, the then Director-General of CSIR, the late Satish Dhawan, Chairman of ISRO and Secretary, Department of Space, Yash Pal, and many others came to see what we were doing and we were thrilled at their extremely positive reaction to the Exhibition.

By the middle of March 1977; the Exhibition became reasonably widely known.

I recall that never before had my whole family (my wife, our daughter and our son whom, we had taken out of school to be with us in Delhi) spent so much time outside of Hyderabad within the country, for any purpose whatsoever. Fortunately, we were given an apartment within the NCERT campus which we furnished and made liveable by borrowing material from various friends in Delhi and by taking some from Hyderabad. It was wonderful to have the entire team stay together for nearly eight weeks (some for a shorter period) in Delhi to put up the Exhibition. The carpenters and the electricians whom we had hired from Delhi worked virtually round the clock under the supervision of Surya Prakash who is, today, one of the leading painters of the country. It took two months to be able to put all the display material together with miles of electrical wiring. We had planned to have the work finished by the third week of March 1977 and then have the exhibition opened by the Prime Minster some time in April 1977.

II. THE THEFT

But we had not reckoned with the events that were to follow. When we arrived in Delhi in January 1977, it was the period of emergency during Indira Gandhi's Prime Ministership, and the elections were round the comer. On 21st March 1977, just the date our exhibition was ready to be inaugurated, Indira Gandhi lost the elections and Morarji Desai was subsequently installed as the Prime Minster. The new Government of India now had a hidden Hindutva agenda which we see fully unveiled today; in fact it was the first time in the country when persons like A N Bose were in an advantageous position and those like Rais Ahmed in an unenviable situation. We were well aware of many panels and statements in the Exhibition which would go against a fundamentalist agenda. In fact, one of the purposes of the Exhibition was to spread scientific temper, reason, objectivity and rationality, and to fight obscurantism.

Events occurred in quick succession after 21st March 1977. Soon afterwards, Rais Ahmed resigned from the Directorship of NCERT. With people like A N Bose now having the run of the field, the future of the Exhibition appeared in jeopardy. We tried to have Morarji Bhai inaugurate the Exhibition through connections in the Parliament, but the NCERT and the new Education Minister would not have any of that. I had, therefore, no option but to lock the exhibition on 21st May 1977 and return to Hyderabad with financial liabilities.

Fortunately, a close and dear cousin of mine, Asha Singh, was working in Bal Bhavan and my relationship with the new Director of Bal Bhavan, Gurbax Singh, was good. I, therefore, with the approval of the Director of Bal Bhavan, gave one of the keys of the Exhibition to Asha so that she could periodically get it cleaned. I had two possibilities in mind. The first possibility was that I continue to persuade the Government of India to have the Exhibition opened and all the pending accounts settled. The other possibility was to find a buyer for the Exhibition who would also provide space for it to be displayed.

The first possibility seemed remote. I, therefore, started looking for a buyer. The first person I contacted was the late Rajni Patel who was then setting up the Nehru Centre in Bombay. However, the Discovery of India Hall at the Nehru Centre (the only place in the Centre that could house the Exhibition) was not ready and he said that while he would be happy to pay for the Exhibition, there is no place where it could be displayed. I then approached the Andhra Pradesh

Government, especially the then Chief Minister, M Channa Reddy. Nayudamma also spoke to Channa Reddy about it. Channa Reddy's Government then started working on the proposal of buying the exhibition from the NCERT and exhibiting it in Hyderabad. He sent the then Education Minister, who later became the Chief Minister of Andhra Pradesh, and the Joint Secretary for Education, Daphne d'Rebello, to Delhi to have a look at the Exhibition. They were impressed, and they recommended to the Andhra Pradesh Government that the Government buy this Exhibition and pay me my dues.

From that time onwards, the Government of Andhra Pradesh—two successive Chief Ministers beginning with Channa Reddy, the Education Minister (Venkatram Reddy), the Education Secretary (Gopalakrishnan), the Finance Secretary (B P R Vithal) and the Joint Secretary for Education (Daphne d'Rebello)—worked hard and expeditiously towards getting the Exhibition to Hyderabad. They planned to allocate (as they later did) about 6,000 sq. ft. of the required built-up space on the first floor of the Hyderabad City Public Library in Afzalgunj, along with the entire area at the rear of the building, for the Exhibition which could be cordoned off for the exclusive use of the Exhibition. But fate (as they say!) had other plans for us.

On Monday, 7th August 1978, 15 months after the Exhibition was locked in Delhi, I received a call from B P R Vithal, the then Principal Finance Secretary to the Government of Andhra Pradesh, saying that the Government had finally issued the order to purchase the Exhibition from the NCERT. I was absolutely thrilled. We felt that the Exhibition could now be displayed permanently at a good place, if not in Delhi then in one of the other major metropolitan cities of the country—that is, Hyderabad. In fact, we could not have asked for anything more; we had the fullest support of the Government of Andhra Pradesh besides, of course, the support of numerous leading personalities of the country and also some from outside the country. Just then—on the same fateful day (7th August 1978)—lightning struck.

I remember that late afternoon vividly. B D Nag Choudhury (the then Scientific Advisor to the Minister for Defence and, later, the Vice Chancellor of Jawaharlal Nehru University at Delhi) came to my office to tell me that, at a meeting of the Council of the Indian National Science Academy (INSA) held at Hyderabad that day, I was elected a Fellow of INSA; he expressed the hope that I would accept the fellowship as he said there were rumours that I may not. I assured him that whatever I do after receiving the invitation will be reasonable and will have a sound basis. (I was delighted to accept the Fellowship of INSA in 1978 but resigned from it in 1984 when I also resigned from the Fellowship of the other two national science academies, on matters of principle.) Just then a call came from my cousin, Asha Singh, in Delhi, which shocked me beyond words. She said that she had been getting the Exhibition cleaned regularly, and it was in excellent shape. The preceding Friday (4th August 1978) , the Director of Bal Bhavan, Gurbax Singh, had asked her for the key, ostensibly to show the Exhibition to some visitors. She said that this was being done regularly as the Exhibition was regarded as a show-piece of science for the country. Therefore, she had no hesitation in giving the key to him. However, when on the following Monday, that is 7th August 1978, she asked for the key back, Gurbax Singh did not give her the key. She instinctively went to the Polish Pavilion, peeped into the large hall where most of the exhibits were located, through some cracks in one of the doors, only to discover that the hall was empty and the Exhibition had been removed. The exhibition had been stolen over the week-end without leaving a trace!

I went to Delhi on 9th August 1978, and met the

Director of Bal Bhavan, and the Joint Secretary of Education (Anjani Dayanand) whom I had known well and who had, in fact, been to my house in Hyderabad. Their entire attitude towards the Exhibition had changed with the change in the Government at Delhi—and they had become very anti-Exhibition and anti-science for the flimsiest of reasons. They were not prepared to tell me how and why the Exhibition was removed and who removed it all without telling me, who was responsible for the Exhibition. I tried to see the new Director of the NCERT, but was unable to see him. I then convened a press conference in Delhi on 11th August 1978 (our 20th wedding anniversary) which was very well attended. On the following two days the newspapers had the stealing of the Exhibition as a lead story; they compared it to the burning of the library at Alexandria by the Romans. The story of the disappearance of this Exhibition was covered by a large number of newspapers and magazines in India and abroad, including *Nature* and *Science*.*

This story, as it unfolded, is told in detail in the next part of this book** (part II) by Vasantha Surya; it is reproduced here as it was written by Vasantha in the late 1970s. Later, when all the pieces of the process of disappearance of the Exhibition were put together, it became obvious that the Exhibition was removed at the orders of the then Minister for Education in the Janata Government (P C Chunder), perhaps at the initiative of the NCERT. Apparently, they did not want it to be shown anywhere. After all, the method of science is all about questioning, and no totalitarian Government wants people to be told that it is their right to question and what the conditions are under which they may exercise this right. The Government of India

---

* See Chapters XIV and XV of the present book.

** The book from which this article is taken.

feared that the opening of the Exhibition could lead to a snowballing effect. The Exhibition, it was felt, argued against the development of the kind of fundamentalist society that they were working towards in the country. The Government wanted us to go back 2500 years in history; the Exhibition impelled you to go forward.

I was able to meet Morarji Desai, the Prime Minister, in connection with the Exhibition, through H Y Sharda Prasad, on 27th September 1978. Morarji Bhai (unlike Mrs. Gandhi, who always welcomed us with a smile) didn't even look up when I walked into his room. I sat down on the chair opposite him in his office and told him the story of the Exhibition. With his head still bent down, he asked me why wasn't he told about the event earlier. When I said that it was all over the newspapers in the country, he replied that he never read newspapers. I then said that it was also discussed on the floor of the Parliament. Fortunately, he did not say that he never went to the Parliament! He appointed an eye-wash committee headed by one Dr. Mathur who was then the Director of the National Institute for Educational Planners and Administrators in the NCERT campus in New Delhi; as expected this one-man committee exonerated the NCERT and the Government of all blame for the fate of the Exhibition. The Indian Rationalist Association (IRA) then (in 1979) filed a Writ Petition which was admitted in the High Court of Andhra Pradesh and eventually heard by a Bench.

Part III of this book* provides copies of the Writ Petition by the IRA and of the affidavits filed by the primary defendants named in the petition (NCERT and me). Part IV of this book* provides some of the documents (including those that were presented to the Court by our lawyer, K G Kannabiran), that relate to the case and the story of the

---

* The book from which this article is taken.

Exhibition as narrated above. (The affidavits include the Mathur Committee Report.) The court judgement is given in Part V of this book.* Part VI* reproduces the contents of a brochure brought out at the time of the opening of the Exhibition at Hyderabad on 12th November 1984; it includes a summary of the content of the Exhibition. The last part of the book* (part VII) has the details of the visuals and the text of the commentary in the film on the Exhibition made by the Films Division of the Government of India.

An annotated version of the text of the Exhibition, fully illustrated, will be published separately.

III. THE RESURRECTION

After the Method of Science Exhibition disappeared from the Bal Bhavan in early August 1978, all the officials concerned, that is, the then Director of NCERT, the Joint Secretary and the Secretary, Education, Government of India, and the Director of Bal Bhavan, were tight-lipped about it so that, initially, no one, including the press could find out what had happened to the Exhibition: whether it had been totally destroyed or it had been shifted from the Bal Bhavan campus. A proper shifting of the Exhibition which had miles of electrical wiring; expensive and sensitive equipment such as microscopes, lasers, VCRs and projectors; several hundred original paintings and drawings; expensively done text panels; material for live shows and experiments; and valuable documents and books; would have needed round-the-clock work by a qualified team over at least a week, if not two. Since the job was done within a few hours (as we

---

* The book from which this article is taken.

had reasons to believe), we feared that the Exhibition would have been badly damaged (this fear was proven to be justified, by subsequent events). Whosoever was responsible for the shifting of the Exhibition (assuming it was not totally destroyed and consigned to a garbage dump) had no sensitivity towards the work of art that the Exhibition was. Later, we had incontrovertible evidence that the Exhibition was shifted by voluntary workers who had an intimate association with Jan Sangh, the major constituent of the ruling coalition at that time. The widely held view in responsible quarters was that this was done by volunteers of the RSS, or those who had sympathies with such organisations. The insensitivity of such people to real works of art is widely known.

What was gratifying was that the Government of Andhra Pradesh (AP) pursued the matter and eventually located the Exhibition in one of the NCERT godowns. During the period between the Exhibition being ready for opening in Delhi (March 1977) and the taking over the Exhibition by the Government of Andhra Pradesh, there were several enquiries from various parts of the world (such as Holland and the USA) about the possibility of the Exhibition being transferred to their country. The A.P. Government negotiated with the NCERT and, finally, in 1980, purchased the Exhibition from the NCERT by paying to the NCERT whatever had been spent by it on the Exhibition till then. The credit for this must be given to Daphne d'Rebello, the then Joint Secretary for Education, and Gopalkrishnan, Secretary, Education, in the Government of Andhra Pradesh, both of whom had also support from the then Minister for Education of AP. The above team and those who followed them in the AP Secretariat, worked hard to negotiate the deal with the NCERT and to have the Exhibition eventually packed and brought from Delhi to Hyderabad. They also paid all the debts that I had accumulated till then. They even offered to pay me a sum of Rs.25,000/- towards my intellectual property right on the Exhibition. The CSIR, of which I was an employee at that time, permitted me to accept this money but I eventually decided against it because it didn't seem to me ethical to do so.

The Government of Andhra Pradesh then left no stones unturned to see that the Exhibition was resurrected. It took some two years to put it back in shape and to set it up on the first floor of the rear side of the Andhra Pradesh State Public Library. The entire first floor (about 6000 sq.ft.) and nearly an acre of land behind it had to be redone for the Exhibition, and the Government of Andhra Pradesh supported us fully in this endeavour and took care of all the expenses. The Andhra Pradesh Science Centre was asked to be the nodal agency to take care of the Exhibition.

The Government of Andhra Pradesh's support at that time towards an activity such as the Method of Science Exhibition must be regarded as unparalleled in Government history. We needed close to 30-40 volunteers every day to run the Exhibition. The Government allowed us to put in an advertisement in the newspapers. Several thousand people applied for becoming volunteers for MOSE. The Government agreed to pay them a reasonable sum per day for working for the Exhibition. Many of the applicants from outside Hyderabad were willing to spend their vacation or holidays to work for the Exhibition. We interviewed some 700 persons and selected several hundred out of them. All the selected ones were given training for all of the nearly 40 positions in the Exhibition. The system that was worked out was as follows. The volunteers would receive a reply-paid post card giving the allotment of dates for them and also the positions that they would be asked to man on a particular date, each position having been numbered. They would then confirm their acceptance on the reply card. All this was done sufficiently ahead of time to allow others to be asked, in case anyone in the first list did not find a date suitable. The system worked extremely well for many months and the Exhibition, I believe, changed the lives of the volunteers that included people of varying professions and avocations, housewives, students belonging to various field of knowledge, people working in administration, and so on.

The Exhibition was to be inaugurated by Indira Gandhi who was by now back in power at Delhi. However, one day I received a call from her in which she said that she was under great pressure to inaugurate this Exhibition and would have very much liked to do so. Unfortunately, she felt that the political situation in Andhra Pradesh would not be conducive to her visit. (At that time N T Rama Rao was the Chief Minister of Andhra Pradesh.) She, therefore, suggested that the Exhibition be inaugurated by a scientist, such as M G K Menon, FRS.

I agreed to this suggestion and Mrs. Gandhi sent a wonderful message for the Exhibition. In fact, that was one of the last messages that she sent before her assassination.

The Exhibition was inaugurated the day following the official mourning period after Mrs. Gandhi's assassination—that is, on 12th November 1984. At the inaugural function, several of the country's high-profile scientists and citizens such as Satish Dhawan, Yash Pal, M G K Menon, Manju Sharma, Rais Ahmed, TN Khoshoo, Bal Phondke, Bakul Patel, and Vasantha Surya were present. The Chief Minister N T Rama Rao, was also present. The inauguration of the Exhibition was widely covered in the national press.

The Exhibition set up as above was filmed by the Films Division of the Government of India. The nearly one-hour film was directed by Mr. Engineer of the Films Division. However, when he started putting everything together they found that there were many gaps. Clearly, the filming had been done without adequate preparation. To fill in the gaps, my colleague, Chandana Chakrabarti, and I visited Bombay over many week-ends when the Film Division kept their office in Bombay open for us to help edit the film. It was interesting for us to sit on the editing table. We discovered many gaps for which visuals had to be found; these gaps were filled using ingenious methods. We then also wrote the commentary. The film was finally released in the late 1980s. At that time it cost Rs.18,000/- per copy on 35-mm. Video cassettes were also prepared although the transfer from 35-mm film to video didn't seem as good as it should have been.

The Exhibition spawned many other activities. A short film called *The Four Steps* was made by Girish Vaidya, Joint Director of the Films Division at that time; this short film was shown in more than 700 picture houses in the country in the late 1980s and also on the national television. Several agencies of the Government concerned with education in New Delhi also made films on the same theme which were broadcast on the national television.

The text of the Exhibition was serialized in *Science Today* in 1980, and also translated into several other languages.

However, even though the Exhibition in Hyderabad was all set to become a major tourist attraction, the input of time required from all of us could not be sustained by us beyond a few months. We had to finally leave everything to the Andhra Pradesh Science Centre. As soon as we did that the interest of the Government of Andhra Pradesh in the Exhibition waned; by this time the Secretary and the Joint Secretary of Education, the Minister of Education and the Chief Minister, had also changed. Under no pressure from us the support to the Exhibition dwindled. The Exhibition had live shows and there were several experiments which were being repeatedly done during the day for the audience; all these were handled by the volunteers selected and trained as mentioned above. The system of volunteers that had worked so well for many months now came to a halt and the Exhibition started deteriorating. Initially, for many months we had the maximum permitted number of visitors (booking had to be made in advance for looking at the Exhibition) each day: this number started falling. This was an outstanding example of the apathy of the new Government to something which could have, if properly maintained, become a major tourist attraction in Hyderabad. With the exhibition falling into disrepair, it became a liability for the Andhra Pradesh Science Centre and the Government. The State Public Library also began to demand back the place that it had given for the Exhibition; this demand could not be ignored as the Exhibition was no longer performing the function it was supposed to perform.

At this point, the Birla Science Centre at Hyderabad approached me for taking the MOSE over and putting it on the first floor of their Science Museum. I was delighted at this idea and encouraged B G Siddharth, the Director of Birla Science Centre, to go ahead with taking over of the Exhibition. The Exhibition was transferred from its location in the State Public Library premises of the Government of Andhra Pradesh, to Birla Science Centre in September 1992.

However, instead of being the beginning of another resurrection of the Exhibition, it was the beginning of an end. The Birla Science Centre never organized the Exhibition properly. It didn't even put the panels in the right order. I suggested to Siddharth several times that, with a small investment, we could make it look even better than what it was in the Andhra Pradesh Public Library. However, nothing was done. Unfortunately, none of us had much time to pursue this matter, but we believed that Siddharth will do something about it. This belief was totally misplaced. The final plundering of the exhibition happened, when Siddharth realizing the value of the original paintings of Laxma Goud and Surya Prakash, decided to pull out only the paintings and add them to the Birlas' art collection while literally dumping the rest of the exhibition. That was the final burial of the MOSE.

IV. THE END AND, THEN, THE NEW BEGINNING

Premanand, the well-known propagator of scientific temper and antagonist of belief in the supernatural and in miracles, who has also been the recipient of the NCSTC's (National Council for Science and Technology Communication's) national award for popularisation of science, and who founded and edits *The Indian Skeptic*, heard me talk about this Exhibition at a meeting in 2000, and said that he would like to take this Exhibition to Kerala. Subsequently he collected some money to put up the Exhibition in Kerala. He then visited Hyderabad and together we went to Birla Science Centre to talk about the possibility of shifting the Exhibition to Kerala. To our great surprise and dismay, I found that there was no Exhibition left at the Birla Science Centre. It was the second destruction of the Exhibition. When we asked Siddharth as to what had happened, we were told that the Exhibition had deteriorated to an extent that it had to be removed. This was a joke as almost all the items in the Exhibition were in perfectly good shape when I had seen it last.

There were microscopes and lasers which could not have become unworkable to an extent that they could not be repaired. There were several hundred lights with shades which could not have become junk, specially in a place like the Birla Science Centre. Clearly, there was a deliberate effort there as well to destroy the Exhibition and one wonders if the purpose of having the Exhibition transferred initially from the Government of Andhra Pradesh to Birla Science Centre, was to eventually destroy the Exhibition respectably. A list of all the exhibits transferred from the Public Library location to Birla Science Centre is available and anyone who examines it is bound to come to the conclusion that most of the items in the Exhibition could not possibly have deteriorated on their own, specially at a place like the Birla Science Centre at Hyderabad, to a stage that they could not be resurrected. This was the second deliberate destruction of what should have been regarded as a national treasure.

The text of the Exhibition and all the documents connected with it, as well as a collection of slides of the Exhibition and the film and the video-tape on it, are still available. With the help of this material, attempts are being made by Premanand, the Editor of this book* to recreate a replica of the original exhibition, as we write this article. We hope that this replica would be created and that the Exhibition would be resurrected the second time. Those who wish to contribute towards this resurrection and the continued replication and maintenance of this Exhibition, may kindly get in touch with Premanand at the following address:**

B Premanand
Editor & Convenor, *Indian Skeptic*
11/7, Chettipalyam Road,
Podanur - 641 023, Tamil Nadu, India

* The book from which this article is taken.

** *Author's note:* As of writing the present book, Premanand has been very ill.

XIV

A REPORT IN *SCIENCE* ON THE METHOD OF SCIENCE EXHIBITION*

Indian Science Exhibit Sits in Limbo

Efforts are under way to locate and reassemble a scientific exhibit from India—dismantled by government agents before the public ever saw it—as a central feature of the United Nations Conference on Science and Technology for Development, to be held in Vienna next August.

The exhibit, which had been sitting in a locked gallery in New Delhi for 14 months, was surreptitiously carted off one night last summer at the behest of its original sponsor, the National Council of Educational Research and Training (NCERT), a quasi-governmental body. It was apparently the victim of the shifting political situation resulting from the election of prime minister Morarji Desai, who defeated Indira Gandhi in March 1977. Conservative politicians evidently took offense at the tone of the exhibit, which challenged beliefs in astrology, "godmen," and traditional homeopathic medicine.

The only information about the exhibit available to *Science* is a sheaf of material sent to an American friend by P M Bhargava, who planned and oversaw construction of the exhibit. Bhargava is an internationally known scientist who heads the Centre for Cellular and Molecular Biology at the regional research laboratory in Hyderabad.

---

* *Science*, 27th April 1979, Vol. 204, p.393.

The exhibit, designed to present "the method of science" to the general public, originated as a result of a suggestion by the then director of NCERT to Bhargava. Largely financed by the Vazir Sultan Tobacco Co. and NCERT, it was a multimedia show featuring the efforts of scientists, engineers, artists, film-makers, and musicians. The exhibit cost about 300,000 rupees to put together but its market value outside India was estimated at $1 million. Constructed in Hyderabad, it was moved to New Delhi in early 1977, where it was set up in a gallery of the Bal Bhavan Society, an organization devoted to studies and welfare of children. It was supposed to be opened by Prime Minister Gandhi but this became infeasible when she lost the election.

Leadership of NCERT subsequently changed and that agency withdrew its support. So the exhibit sat in the closed gallery while Bhargava went around looking for new backers and a new home for it. Then one August night 100 men with trucks appeared and carried everything away in a matter of 6 hours—very likely causing extensive damage, since to pack the exhibit properly would have required several weeks.

The new head of NCERT, SK Mitra, explained to the press, which gave considerable coverage to the episode, that the exhibit had been taken down because there were scientific controversies about its content. Mitra was reported as saying "a group of scientists" objected to certain displays, including a surrealistic picture of a reclining nude dreaming—designed to portray the nature of tachyons, particles that travel faster than light. The "scientists," however, were not named; several prominent ones, who were, have reportedly spoken highly of the exhibit.

Several displays reportedly offended political sensibilities. One panel in the exhibit was constructed as an attack on Indian "godmen." A *sadhu* (religious leader) was shown materializing objects from the air and the viewer was invited to judge whether that squared with the first law of thermodynamics. The display also emphasized that "science has no high priests."

Another part of the show cast aspersions on astrology by displaying samples of varying predictions for the same week contained in different magazines, and asking how these predictions withstood scientific scrutiny. Elsewhere, the exhibit questioned whether the "method of science" was applied in the preparation and use of certain salt pills in homeopathic medicine.

Another offensive section, related to the history of science, featured Marx and Lenin as pioneers in applying scientific principles to social theories and economic planning.

The exhibit ended with a quote from the out-of-favour Mrs. Gandhi to the effect that "we want scientific thinking to destroy superstition which has darkened our lives."

Bhargava believes that NCERT moved in to snatch the exhibit when officials heard of proposals to put it on display in Vienna. An Indian scientist in this country told *Science* that the move was very likely initiated by "henchmen" eager to please Prime Minister Desai—"you know that our prime minister drinks urine," he said, a practice that falls in line with belief in homeopathic and naturopathic remedies. Bhargava is said to have met with Desai to plead for release of the exhibit, but so far no action has been taken.

The Rationalists Association of India, one of many organizations devoted to promoting "scientific temper" in the country, has gotten up in arms over the handling of the exhibit. One member wrote a letter to *The Times of India* complaining that the country was "sinking deeper and deeper into superstition, fatalism and religious hypocrisy." He related that one scientist had been forced to resign as Vice-Chancellor of Bangalore University after he asked prominent "godman" to subject miraculous performances to a scientific probe. Last December the Rationalists Association filed a writ claiming that NCERT's dismantling of the exhibit was a violation of the Constitution. This is said to be the first legal case of its kind in the country.

The exhibit would have been the first of its kind designed to acquaint Indians with scientific thinking. The overall objective, according to Bhargava, "was to show how useful and important it is to make the method of science an integral part of one's thinking and living, and how one can use this method profitably to solve one's day-to-day problems." To a Westerner it sounds harmless enough, but the designers of the exhibit apparently underestimated the degree to which advocacy of doubt and questioning could be construed as a threat to government authority.

In Indian press accounts of the affair, a professor is quoted as explaining the government's objections: "we have no tradition of genuine doubt in our philosophy. One can either accept, reject, or remain passive, but one may not doubt or enquire. Doubt, even in the West, dates clearly only from Roger Bacon's time."

Ward Morehouse, president of the Council of International and Public Affairs in New York, was vastly impressed with the exhibit and thought it would be a great way of showing that "science has very much come of age in at least some Third World countries in the past 30 years. This message, to be fully convincing, must come from the Third World itself." But whether it will come from the "world's largest democracy" is very much in question.

XV

A REPORT IN *NATURE* ON THE METHOD OF SCIENCE EXHIBITION*

Science for Schools: Making an Exhibition of Itself

India's "Method of Science" exhibition, first planned in 1975, will probably not now be opened until after the general election expected within the year. The exhibition, intended for senior secondary school students, was planned as a permanent display, presenting science not as a collection of achievements and gadgets but as an intellectual discipline based on deductive reasoning and inference. It was assembled in New Delhi in 1977, dismantled overnight by government order in August 1978, reassembled in Hyderabad in 1983 but is still, at the time of writing, not open to the public.

The exhibition was originally proposed by Dr Rais Ahmed, the then director of the National Council of Educational Research and Training (NCERT) and designed by Dr Pushpa Bhargava, director of the Hyderabad Centre for Cellular and Molecular Biology (CCMB). The completed design was opposed by moral reformers, leaders of religious cults, homoeopaths, civil servants, accountants, politicians and even some members of the scientific establishment who, with the change of government in March 1977, tempered their original enthusiasm to the prevailing official line. The

---

* *Nature*, 12th April 1984, Vol.308, p.598.

Indian Humanist Society, however, took legal action on behalf of the exhibition, and the issue was taken up in the standing committee on the Safeguard of the Pursuit of Science of the International Council of Scientific Unions.

Most overt criticism has been detailed. The financial lobby questioned Bhargava's spending on visual material ("Could not the captions have been written with chalk on blackboards?"). Another controversy arose over a display allowing visitors to decide whether propositions are true, false or not yet determined. A wraith-like figure, identifiable as female only by its floating hair, which simultaneously leaves and enters a door is used to illustrate the concept of motion faster than light, but was condemned by some as a lascivious nude. Others objected to the inclusion of Marx, Engels and Lenin in the series of "21 landmarks in the history of the method of science".

An exhibit entitled "Science has no high priests who cannot be questioned" contrasted a panel showing a young scientist at a conference questioning the results of an eminent lecturer with a "godman" surrounded by a group of humble disciples. To avoid accusations of caricature, the face of the "godman" was a portrait of one of the organizers, but controversy sprang up on the ground that the gold watch in his hand might remind viewers that materializing watches is part of the repertoire of a particular "godman" with a large following. (The artist obligingly replaced the watch by a ring, but the critics were still not satisfied.)

The more serious objections have never been made explicit. The exhibition questions the value of homoeopathic medicine, without which the Indian health services would not function. It also questions astrology, while several leading politicians are known to order their professional lives according to the stars. Perhaps most dangerous of all, to the Janata Government of 1977-79, it is anti-authoritarian.

It teaches nothing and asks visitors to accept nothing on trust, even the message of the exhibition itself.

Official support for the exhibition lapsed with the defeat of Mrs. Indira Gandhi's government in 1977. Ironically, it might well have opened in Delhi if Dr Rais Ahmed and the Education Minister, Professor Nurul Hasan, had not pressed for Mrs. Gandhi herself to inaugurate the exhibition. The approach of the elections, however, meant that Mrs. Gandhi had no convenient slot in her diary until too late.

Recently, political events have once more conspired to delay the opening. The Hyderabad authorities, who have given their backing to the reassembled exhibition, felt that it would be proper for the opening ceremony to be performed by Professor Nurul Hasan, who is now Indian Ambassador to the Soviet Union. Plans for him to open the exhibition in March were disrupted by the death of Mr. Yuri Andropov. The inauguration is now tentatively scheduled for November. While the exhibition itself stresses the anti-authoritarian nature of science, those responsible for it seem unwilling to do without the blessing of a political guru.

XVI

A STATEMENT ON SCIENTIFIC TEMPER*

INTRODUCTION

Nehru Centre, which owes its inspiration to the great ideals of Pandit Jawaharlal Nehru, the architect of modern India, has been steadily developing itself into an institution of national and international importance. Its main objectives include, among others, inculcation and promotion of new social values, secularism, national integration and self-reliance as well as rational outlook on life.

The nation owes a deep debt of gratitude to Jawaharlal Nehru, more than to any other, for the sustained growth and many-sided development of modern science and technology in India, as viable instruments of social transformation. The need of the time is the diffusion of science and technology into the societal fabric at all levels.

---

* This statement was an outcome of a meeting organised by the Nehru Centre of Bombay, at the Coonoor Club, Coonoor, from 22 to 25th October 1980. The meeting was chaired by Mr. P N Haksar, and convened jointly by Mrs. Bakul Patel (Trustee and Member, Executive Committee, Nehru Centre), Dr P M Bhargava (Head, Centre for Cellular and Molecular Biology, Regional Research Laboratory, Hyderabad), and Dr B V Subbarayappa (Director, Discovery of India Project, Nehru Centre).

At that time, Mr. A R Antulay was the Chairman of the Nehru Centre; Dr R Ramanna, the General Secretary; Mr. Ram Batra, the Treasurer; and Mr. Ajit Mehta, the Joint Secretary.

This can only be achieved by promotion of what Jawaharlal Nehru chose to call THE SCIENTIFIC TEMPER—a rational attitude, the importance of which he emphasized time and again. Indeed, the Scientific Temper has to be fostered with care at the individual, institutional, social and political levels.

In view of this, Nehru Centre thoughtfully organized a Group Meeting of scientists, sociologists, historians and others, at Coonoor in October last year to prepare a statement on the definition and promotion of Scientific Temper. It is the fervent hope of Nehru Centre that this Statement will pave the way for more intensive efforts in the direction of promoting scientific attitude among our people in all walks of life.

On behalf of Nehru Centre, I would like to thank all those who extended their valuable cooperation in organizing the Group Meeting and the preparation of the Statement in all its aspects, specially the first General Secretary of Nehru Centre, Shri Rajni Patel.

Bombay, RAJA RAMANNA
July 10, 1981 General Secretary

FOREWORD

The Nehru Centre arranged for some of us to assemble together in a quiet corner of our country to share our common concern at the accelerating pace of retreat from reason. The venue of our meeting was Coonoor, so lush green and full of promise as our entire land is.

For four days and nights, from October 22–25, 1980, we discussed and debated what needed to be done to halt the process of decay of reason and rationality. I had the honour of presiding over the deliberations. The end result of it all was a Statement on Scientific Temper.

The Statement was subsequently shown to others, it was further refined. We now present this Statement. We are not unaware of its inadequacies. However, it is our earnest hope that the Statement will generate a wider debate and discussion in our country.

There are more than two million scientists and technologists in our country. In addition, we have a large number of economists, historians, sociologists and anthropologists, lawyers, doctors, administrators, management specialists and teachers who, in one way or another, apply the scientific temper and scientific methodology in pursuit of their respective professions and disciplines.

If the Statement succeeds in generating a nation-wide discussion, it will also, hopefully, generate a movement for the much needed second renaissance in our country. The first renaissance inspired the struggle for freedom. The second must of necessity provide the necessary fillip for the restructuring of our country embodying the aspirations of our people.

Only in the measure we succeed in installing scientific temper as the dominant ethos of our collective being, can we hope to face the accumulating problems of our national existence. We must understand that it is not going to be easy. We shall have to do a great deal of heart searching ourselves.

It is often argued, with seeming profundity, that while scientific temper is alright, it does not satisfy humanity's spiritual needs; that the entire realm of art and music, poetry and drama fall outside its ambit. In answer to such critics, I can do no more than remind ourselves of how Jawaharlal Nehru resolved the seeming contradiction between our material and spiritual needs.

In the *Discovery of India*, he defines in the following terms his own attitude:

> The real problems for me remain problems of individual and social life, of harmonious living, of a proper balancing of an individual's inner and outer life, of an adjustment of the relations between individuals and between groups, of a continuous becoming something better and higher, of social development, of the ceaseless adventure of man.
>
> In the solution of these problems the way of observation and precise knowledge and deliberate reasoning, according to the method of science, must be followed. This method may not always be applicable in our quest of truth, for art and poetry and certain psychic experiences seem to belong to a different order of things and elude the objective of science.
>
> Let us, therefore, not rule out intuition and other methods of sensing truth and reality. They are necessary even for the purposes of science. But always we must hold to our anchor of precise knowledge tested by reason... we must beware of losing ourselves in a sea of speculation unconnected with the day-to-day problems of life and the needs of men and women. A living philosophy must answer the problems of today.

4/9 Shanti Niketan P N HAKSAR
New Delhi – 110021
February 17, 1981

PREAMBLE

The history of humanity bears witness to periods of enlightenment as well as to periods of darkness. It bears witness to the rise and fall of civilizations. Through all the vicissitudes of time the knowledge gained by humanity has retained a quality of indestructibility. Viewing the entire panorama of the universal history of mankind, one becomes conscious of a continuous but forward movement towards greater knowledge and to an increasing capacity of human beings to exercise control over their environment.

While humanity as a whole accumulates knowledge, there is no guarantee that the availability of such knowledge will, by itself, enable every country to use it successfully for its own advancement and the well being of its people. There are examples in history where predominant social, political, cultural and value systems inhibited the absorption of knowledge, resulting in periods of stagnation, decay and retreat from reason, rationality and science. Though the Renaissance began in Italy, and Galileo, the harbinger of modern science, was an Italian, adherence to obscurantism enforced by the Church led Italy to losing the benefit of Renaissance which fertilized northern parts of Europe. The Renaissance and the Reformation then combined together to revolutionize thought as well as society—in northern Europe first and in Italy only later.

In our country too we have known of periods of creativity when the spirit of enquiry led to the accumulation of scientific knowledge; there was creativity in literature, music, arts and crafts. However, we have also known of periods when the spirit of enquiry got extinguished. During those long stretches of time everything was reduced to unquestioning dogmas and to the performance of dead rituals. There was deadening of curiosity and questioning. There was only passivity and acceptance. And finally, we were overtaken by the greatest of disasters—our complete colonization and subjugation to British imperialism.

Contemplating our decline, decay and subjugation, some of our best minds began asking themselves why and how it all happened. This spirit of enquiry and questioning gave birth to a wide social cultural movement, which we call the Indian renaissance. The best Indian minds in the pre-independence times insistently propagated the need for the people to think independently and fearlessly, and to question traditional beliefs. This effort, in time, produced a critique of the colonial system. Out of this critique was born a powerful national movement for our liberation. The British imperial system, aligning itself with the vested interests, endeavoured to counter the broad stream of nationalism by encouraging revivalism and obscurantism. And though Indian renaissance never elaborated a critique of our entire ancient society and unfortunately made compromises, the urge to acquire knowledge and a scientific outlook remained strong. The spirit of questioning ultimately overwhelmed an imperial system which seemed so powerful and even immutable.

There is a wide awareness in our times that we are living in a scientific age of great discoveries in science, affecting and moulding both our material and social existence. It is indeed remarkable how a comparatively small number of physical laws seem sufficient to explain a great part of behaviour of matter, right from the huge and massive heavenly objects located at the very edges of outer universe to the minute regions of atoms and atomic nucleus. In life sciences, we are in the midst of far reaching, even revolutionary changes. The entire history of humanity shows that it is the scientific temper which not only created and promoted science, but also gave humanity the means to effect the natural and social environment. It is, therefore, the scientific temper which is the most precious heritage of humanity. It is the result of incessant human labour, search and struggle.

Jawaharlal Nehru gave an impetus to scientific temper by setting before the people the target of catching up with the rest of the world with the help of science and technology. He unfolded the perspective of leap-frogging the centuries. Implicit in such a vision was a vast change in the intellectual climate of our people. Our Constitution and subsequent Resolution on Science Policy were predicated upon the assumption that our ancient society needed basic changes. However, there was not enough appreciation of the relationship between the objectives to be achieved and the methods as well as the instrumentalities appropriate for bringing about the desired changes. No systematic and sustained effort was made to work out, specifically and concretely, what needed to be done to build a society which is animated by a spirit of enquiry rather than passivity and acceptance. The result of this lack of directed efforts was accommodation, even compromise, with the forces of obscurantism and with the existing inegalitarian social and economic structures. Failure to give mass dimensions and appropriate institutional forms to scientific temper, more specially in our educational system, led to the erosion of confidence in our capacity to mould our destiny.

In such an environment, scientific temper has been beleaguered and besieged by deep-rooted structures of an ancient society with superimposed colonial structures. Consequently, there has been frustration of our hopes of optimising the results of the application of science and technology for our national reconstruction. Inevitably, such frustration has encouraged a search for and reliance upon authority. Inevitably too, there has been a growth of tendencies to escape into magical beliefs and instant solutions. Even science and technology are being offered not as methods of enquiry or value systems but as magical cures for our ills, reminding one of the time when Roman intellectuals sought refuge in Levantine magic. There is inadequate appreciation of the close interaction between science and technology and society, and of the fact that the benefits of science and technology can reach the people only if the socio-economic conditions are conducive. If the cultural environment, socio-economic conditions and institutional structures inhibit the spirit of enquiry, the desired results can never be achieved.

The gravity of our predicament is increasing day by day. While we rank high among the industrialised countries in the world and are the third largest country in the world in regard to the stock of manpower trained in science and technology, we are close at the bottom of the list in terms of per capita food consumption, longevity, health care and general quality of life. We have all the technology available right now within the country to give water, food, shelter, and basic health care to our millions. And yet we do not. Something has gone wrong. The logic of planning and the logic of our socio-economic structure are at variance. Hence, our failures and disappointments.

In such an environment, there is an erosion of belief in the capacity of human faculties to solve national problems through a systematic critique of the existing social situation. There is a cancerous growth of superstition at all levels. Rituals of the most bizarre kind are frequently performed often with official patronage. Obscurantist social customs are followed even by those whose profession is the pursuit of scientific enquiry. Our entire educational system works in an atmosphere of conformity, non-questioning and obedience to authority. Quoting authority of one kind or another substitutes enquiry, questioning and thought.

Obscurantism and irrationalism practiced by a hierarchy of authorities, has the predictable effect of reinforcing retreat from reason. Voices raised against such a state of affairs get silenced. The decision-making processes are increasingly being divorced from any rational purpose or design. There is no long-term perspective based on ascertained facts and scientific analysis. Adhocism, whims and the narrowest of considerations take the place of well-planned programmes. Priorities, if any, are fixed without sufficient data-base and without any attempt at scientific evaluation of national needs, potentialities and feasibility of implementation. Mere slogans tend to be used as substitute for action and for creating an illusion of achievement. Dramatic crash programmes are launched. These, inevitably, crash. There are no perspective plans. Even Five Years Plans have been reduced to annual exercise of allocating funds.

As our country enters the last two decades of the 20th century, the need to move forward is becoming ever more insistent. We either overcome the obstacles or we shall be overcome by unreason and dark reaction. We must understand the meaning as well as the imperatives of scientific temper, representing as it does, humanity's assertion of being in charge of its destiny and not a passive victim of malevolence of stars. To do so, we need to actively combat beliefs which erode scientific temper and undermine its growth. Only then shall we illumine our darkening national horizon and provide our people, once again, with a vision and a method for translating that vision into reality. Such a vision must have scientific temper as its integrating bond.

ATTRIBUTES OF SCIENTIFIC TEMPER

Spread of scientific temper in society is much more than the spread of science or technology. Scientific temper is neither a collection of knowledge or facts, although it promotes such knowledge; nor is it rationalism although it promotes rational thinking. It is something more. It is an attitude of mind which calls for a particular outlook and pattern of behaviour. It is of universal applicability and has to permeate through our society as the dominant value system powerfully influencing the way we think and approach our problems—political, social, economic, cultural and educational.

Scientific temper involves the acceptance, amongst others, of the following premises:

a) that the method of science provides a viable method of acquiring knowledge:
b) that human problems can be understood and solved in terms of knowledge gained through the application of the method of science;
c) that the fullest use of the method of science in everyday life and in every aspect of human endeavour—from ethics to politics and economics—is essential for ensuring human survival and progress; and
d) that one should accept knowledge gained through the application of the method of science as the closest approximation to truth at that time, and question what is incompatible with such knowledge; and that one should from time to time reexamine the basic foundations of contemporary knowledge.

The method of science, therefore, constitutes a regenerative process for collecting information and processing the collected information to create meaningful patterns leading to an ordered understanding of nature of man himself, his natural and social environment. In this sense, the method of science encompasses all aspects of communicable human knowledge and cuts across all artificial compartmentalization like natural science, social science, applied science, etc.

The spirit of inquiry and the acceptance of the right to question and be questioned are fundamental to scientific temper. It calls upon one to ask the 'how', the 'what', and the 'why' of an object, event or phenomenon. It further calls upon one to exercise the right to question, provided of course, the questioning of an existing theory, hypothesis or statement or social situation is done in accordance with the scientific method and is not merely a bare assertion of one's belief. Scientific temper is, therefore, incompatible with the acceptance of authorities of all kinds or of 'high priests' who may not be questioned. It leads to the realization that events occur as a result of interplay of understandable and describable natural and social forces and not because someone, however great, so ordained them. These forces are often complex and intertwined and have to be analytically disentangled.

Scientific temper is compatible with observation and insight, reasoning and intuition, systematic work and creative impulse. It gives rise to an attitude of mind which while being conscious of vast areas of ignorance, is nevertheless, optimistic about human ability to gradually unravel the mysteries that surround us. In this process, scientific temper becomes a part of a culture, a philosophy, and a way of life which leads to pursuit of truth without prejudgement.

Scientific temper implies the recognition that knowledge often progresses by disproving earlier ideas, beliefs, theories and laws. It considers knowledge as open-ended and ever-evolving. It lays emphasis on verifiability and repeatability, wherever possible, and on the fact that scientific theories, laws and fact allow one to make predictions which can be tested. It recognizes that answers to many questions that may be asked at any given time, may not be available at that time. It, then, demands the courage and humility to say, 'I do not know'.

Scientific temper calls for recognition of the several major differences between the scientific attitude and the theological and metaphysical attitude, specially in respect of dogmas proclaimed in the same religion. There is, in fact, essential incompatibility of all dogmas with science. While science is universal, established religions and religious dogmas are divisive. Consider the divisions which exist between Christian, Islamic, Buddhistic and Hindu denominations. Science, in contrast, transcends divisions and is universal.

Scientific temper has deep emotional content and has, within it, a sense of beauty. That is why considerations based on beauty and simplicity have been often invoked to choose between alternative theories that are otherwise equally tenable.

Inherent in scientific temper is a system of value judgements. The inculcation of scientific temper in our society would result in our people becoming rational and objective, thereby generating a climate favouring an egalitarian, democratic, secular and universalist outlook. Consequently, scientific temper cannot flourish in a grossly inegalitarian society where 50% of the population lives below the poverty line and almost 70% of our people, especially females, are illiterate. Social justice, widespread education and unrestricted communication are, therefore, pre-requisites for spread of scientific temper and for optimising the results of science and technology.

ROLE OF SCIENTIFIC TEMPER

Having outlined the essential elements of scientific temper, let us survey our national scene. Despite Jawaharlal Nehru's advocacy of scientific temper, we are witnessing a phenomenal growth of superstitious beliefs and obscurantist practices. The influence of a variety of godmen and miracle makers is increasing alarmingly. The modern tools of propaganda and communication are being used to give an impression that there exist instant and magical solutions for the problems that confront our people.

In an age when man has travelled to the moon and returned safely, astrological predictions based on the movements of planets, or on the lines of one's palm or the number of alphabets in one's name, are widely believed. Food fads and irrational health practices are on the increase. In a poor country where millions live below the poverty line, vast amount of wealth is consigned to *havanas* and *yagnas*.

Myths are created about our past. The origin and role of the caste system is explained in a way that would justify it and imply that some castes are inherently superior. The ancient period of our history is interpreted to inculcate chauvinism which is false pride; the medieval period is misinterpreted in a way that would fan communalism; and the struggle of our people for freedom is over-simplified as if it was the handiwork of a few great leaders and the masses of our people did not matter.

While it is important to understand the origin of these unscientific beliefs, the more immediate and pressing problem is to understand the remarkable phenomenon of their persistence and the resulting social consequences.

The sustenance of such beliefs and superstitions must be recognised primarily as a historical and social process. Such beliefs continue, because they have ready relevance to the personal situations of the majority of our people. Vast uncertainties of our daily lives, frustration of hopes and aspirations of millions, denial of any vision which would sustain the spirit, drives millions to seek mental equilibrium in faith healing. Thus, when one believes that one's miserable personal situation cannot be improved, acceptance of fatalism becomes natural. Beliefs then rationalise the status quo and breed fatalistic doctrines. In such a situation of social and cultural malaise, a major role of scientific temper is to revive confidence and hope and to dispel fatalistic outlook. The campaign to promote scientific temper must inculcate values like equality and dignity of all human beings, distributive justice, dignity of labour, and social accountability of one's actions. All these are essential for bringing about social, economic and cultural transformation of our country.

The emphasis on the method of science does not imply that science and technology have solutions to all human problems at any given time. Indeed, scientific temper warns one against the simplistic view that through the introduction and pursuit of science and technology, most social problems and contradictions will automatically get resolved. The role of reason is to apply scientific knowledge to problems, to grapple with them through the method of scientific inquiry, and to work for social transformation inspired by scientific temper.

We must equally combat the tendency to treat science and technology as a sort of magic. It should be explained that it is unscientific to believe that if scientific and technological solutions exist to a range of problems, these will be automatically adopted. The nature of social stratification and the power structure in a society prevents the acceptance of such solutions. Technologically, one may be able to grow enough food for everyone, but the pattern of income distribution prevents the benefits of increased food production reaching large segments of the population. When the social structure and stratification prevent the application of rational and scientifically proven solutions, the role of scientific temper is to lay bare the anatomy of such social barriers.

If we have to regain our place in the world and are not to be relegated once again to the dustbin of history; if we wish to offer a life of fulfilment to our destitute millions; indeed, if the light of our civilization is not to be extinguished, we have to undertake, on a priority basis, the task of nurturing scientific temper. All of us scientists, technologists, social scientists, educationists, teachers, media men have to join hands and undertake this task. We draw inspiration from the way our people in all walks of life joined hands and struggled against colonial domination of our land and of our minds. We believe it can be done again if only we have the will. And it must be done without any loss of time. Our nation's survival and its future depends on upholding scientific temper. Superstition shall not pass and darken our portals.

Participants in the Coonoor Meeting and Signatories to Statement

Prof. Amit Bhaduri
Jawaharlal Nehru University
New Delhi

Dr. P.M. Bhargava
Centre for Cellular & Molecular Biology
Regional Research Laboratory
Hyderabad

Prof. Bipan Chandra
Jawaharlal Nehru University
New Delhi

Prof. V.K. Damodaran
Regional Engineering College
Calicut

Mr. P.N. Haksar
New Delhi

Mr. V. G. Kulkarni
Tata Institute of Fundamental Research
Bombay

Dr. Dinesh Mohan
Indian Institute of Management
Bangalore

Dr M. V. N. Nair
Indian Institute of Management
Bangalore

Prof. R. Narasimha
Indian Institute of Science
Bangalore

Prof H. Narasimhaiah
The National Educational Society of Karnataka
Bangalore

Mrs Bakul Patel
Nehru Center
Bombay

Mr. Rajni Patel
Nehru Centre
Bombay

Mr. P.K. Ravindranath
Nehru Centre
Bombay

Mr. Mohit Sen
Communist Party of India
New Delhi

Dr B. V. Subbarayappa
Nehru Centre
Bombay

Other Signatories to the Statement
Mrs. Tara Ali Baig
International Union for Child Welfare
New Delhi

Mr. Shyam Benegal
Bombay

Dr. Satish Dhawan
Indian Institute of Science
Bangalore

Prof. Y. Nayudamma
Jawaharlal Nehru University
New Delhi

Mr Ashok Parthasarathi
Electronics Commission
New Delhi

Dr. K. N. Raj
Centre for Development Studies
Trivandrum

Dr. R. Ramanna
Bhabha Atomic Research Centre
Bombay

Dr. S. Ramaseshan
Indian Institute of Science
Bangalore

Prof. C. N. R. Rao
Indian Institute of Science
Bangalore

Dr. A. K. N. Reddy
ASTRA, Indian Institute of Science
Bangalore

Dr. Anand Sarabhai
BIOCENTER
Ahmedabad

Prof. B. M. Udgaonkar
Tata Institute of Fundamental Research
Bombay

All the above signatories signed the statement in their personal capacity.*

* *Author's Note*: Subsequently, after the publication of the above statement, several other well-known academicians, such as Prof. Yash Pal and Dr Rais Ahmed, also signed the statement.

XVII

WHY THE STATEMENT ON SCIENTIFIC TEMPER*

P M Bhargava

There are three main reasons that prompted me to accept the invitation of the Nehru Centre to arrange the Coonoor meeting in October 1980 which led to the Statement on Scientific Temper of which I was one of the signatories.

(a) In my reckoning, there are three *primary* problems that face the country today: education, water and energy. I call them primary problems because unless these problems are solved, all other problems that we talk about, such as population control, health and employment, simply cannot be solved. To equip ourselves to solve the plethora of problems that surround us, it is absolutely necessary that the three primary problems are solved *adequately*. Any adequate solution of the problems of education must involve emphasis on scientific temper—not only as an integral part of the strategy for the process of education, but also as an objective of education and of the value system towards which the education is directed. I have argued elsewhere (Bhargava, *New Quest*, Vol. 15, May-June 1979, pp. 147-158) that education cannot be value-free and, therefore, it is not only desirable but necessary to orient it deliberately towards a set of values

---

* This article appeared in *Secular Democracy*, 1981, Vol. XVI (ii), pp.29-31.

which we accept axiomatically and which have a universality and the sanction of experience through the ages.

(b) We, today, are living in an environment where most decisions taken show little respect for facts or for considerations of objectivity, fairness and justice, leave aside commitment to values such as secularism, socialism and democracy to which, indeed, we pay nothing more than lip-service. This holds true as much for politicians as for the majority of scientists and other intellectuals. Therefore, something needs to be said about scientific temper, which calls for respect for facts, objectivity, reason and the set of values that I have referred to above.

(c) Science and technology are an integral part of our life style today in every part of the world. However, they are presented to us, as of now, as nothing more than gadgetry, whereas it is the temper of science that is much more important as that alone can enable man to make decisions (including those about the gadgetry) conceived in reason—decisions that would bring maximum benefit to maximum number of people.

What Should be Done?

I am giving below what, in my opinion, could be and should be done to help scientific temper prevail in our country:

1) Promotion of scientific temper is now, according to our constitution, one of our duties as citizens. Some examples of what needs to be done in the present context are, therefore, given in the following paragraphs. These examples are only indicative and not exhaustive. A much more broad-based action is necessary as the conflict between current beliefs and activities in the country on the one hand and scientific temper on the other, exists on a wide scale.

2) Education of the right kind—that is, commensurate with scientific temper—must be considered as the single most important tool for the inculcation of scientific temper on the national scale. For this to happen, a continuous process of reviewing of text books prescribed in the country for all levels of school and college education, for unscientific beliefs or statements that are manifestly incompatible with existing knowledge, must be instituted. A separate allocation of time should be made in the school curriculum for all classes, for conveying the meaning, the attributes, the importance and the uses of scientific temper. Every opportunity must be taken to talk about scientific temper directly or indirectly, in text books. A book found to contain material which goes against the tenets of scientific temper should be suitably amended or, otherwise, withdrawn. A part of the responsibility for these actions would no doubt be that of the Government, but the members of public could play a crucial role in ensuring that the above actions are actually taken.

3) The professional societies and the national academies and councils in the country, should not condone or indulge in activities that are not in consonance with scientific temper. They should also actively contradict—for example, through publication of authoritative reports—non-scientific beliefs that gain currency from time to time. State and local academies and academic institutions, including voluntary organizations, should undertake a massive programme of taking science to, and developing a scientific attitude in, people in their respective geographical area, as has been, for example, done with notable success by Kerala Sastra Sahitya Parishad.

These organizations should also engage in mass publication programmes aimed towards development of scientific temper. Periodicals brought out for this purpose by non-governmental organizations should be supported financially, both by the Government and by others, through, for example, grants, voluntary donations, advertisements, etc.

4) As school teachers would have to play a key-role in the development of scientific temper on the national scale, and in ensuring that scientific temper becomes an integral and indelible part of the fabric of our every-day life, suitable material for the school teachers, on scientific temper, must be produced on a mass scale in various languages. School teachers—as well as others—who should wish to spend their time in bonafide activities that would promote scientific temper, must be encouraged to do so, for example, by the granting of special casual leave. Organisations such as the National Book Trust and the National Children's Book Trust should bring out books which would directly promote scientific temper.

5) Professional persons, public leaders, scientists and other academicians must not only exhibit scientific temper in their professional life but also in their personal life; the dangers of dichotomy in this regard, in terms of its effect on the public, must be recognized. Persons in the above category, especially, must begin to feel internally compelled to object to unscientific activities or reports—for example, through letters to newspapers, articles, etc. They must not be daunted by the discouragement that they are likely to meet at the hands of the press in the course of such efforts.

6) The artists in the country should produce works of art, including cartoons, which would help promote scientific temper. The same would be true of film makers in both the private and the public sector. The Films Division of the Government of India should produce documentary films, and the progressive film producers, feature films,

which would centre around a theme related to scientific temper. The enlightened members of the community should bring it to the notice of the public (and of the others concerned) when a film clearly perpetrates a non-scientific belief.

7) Newspapers must check and verify facts before publishing a report, not only on science but also in other areas. They should be particularly thorough in doing so in regard to news items which would go against the temper of science. Investigative reporting in regard to scientific matters must be encouraged, and should become the rule rather than the exception. The major newspapers should have a regular column devoted to the advancement of scientific temper. At least a few major universities in India should have a course on science journalism with emphasis on the development of scientific temper.

8) The All India Radio and Doordarshan should constitute advisory committees for their various stations, consisting of those whose commitment to scientific temper is unquestioned. These committees should advise the radio or the television station of the kind of programmes that would promote scientific temper; a certain percentage of the time available on the broadcasting station should be reserved for such programmes.

9) Science museums should be established on a large scale—on the national, the State, the district and the city levels. The emphasis in these museums should be not on the facts of science but on the spirit of science, on its conceptional framework and on scientific temper.

10) Similarly, science centres should be established around the country which may serve as resource centres for development of material and expertise that could be used

on a mass scale in the geographical area that they cover, and which would directly or indirectly assist in the propagation of scientific temper.

11) Due attention must be paid to the development of languages to take care of the requirement of communication in relation to scientific temper.

12) Manifestos that would be compatible with and encourage scientific temper should be prepared for the village administration.

13) The premises of all schools and colleges in the country—Governmental or non-Governmental—should be available free of cost to all organizations for bonafide activities which would encourage scientific temper.

14) Periodical conference must be held under suitable auspices, bringing together media people, administrators, natural scientists, social scientists, and others who may be concerned with the promotion of scientific temper. Organisations engaged in such activities must establish linkages with one another.

15) A suitable machinery should be instituted for discussion of a proposed new legislation, by experts; and a suitable and reliable database must be created to help arrive at decisions and make statements. Statements should be made by leaders with care and after due scrutiny; they should be compatible with scientific temper and based on facts. The process of decision-making at various levels—from the individual to the national—should be commensurate with scientific temper.

16) Similarly, the stated objectives at all levels, should be in consonance with scientific temper; in other words, they should be reasonable and achievable. The

statement of an objective should, in fact, be accompanied by a statement of the methodology of achieving the objective.

17) The assessment of policies and decisions of others, at all levels, must be made through a process and in a manner that would be commensurate with scientific temper. Alongside, an equally rigid scrutiny of one's own views should become the rule.

18) Maximal use should be made by all concerned, of existing legislation to fight obscurantist beliefs and unscientific claims.

XVIII

THE SCIENTIFIC TEMPER AND THE SCIENTIFIC METHOD IN SCIENCE IN INDIA THROUGH HISTORY, WITH SPECIAL REFERENCE TO BIOLOGY*

P M Bhargava and *Chandana Chakrabarti*

I. INTRODUCTION

There are two ingredients of this presentation: Scientific Temper and History. The importance of looking at history does not need to be emphasized, but we must explain as to why we have chosen to write on scientific temper in the perspective of history as this is not commonly done by historians.

We believe that one of the major objectives of studying history is to help in providing a sound basis for development. All development takes place against the background of history. Scientific temper is an essential tool for development, be it individual development or collective development at various levels—at the level of the

---

* This article was published as an occasional paper no.22 of the project on *History of Indian Science, Philosophy and Culture*, New Delhi, India (31 pages) and also in *Science Philosophy and Culture: Multidisciplinary Exploration*, part one, Ed D.P. Chattopadhyay and Ravinder Kumar, published by PHISPC / ICPR, New Delhi, 1996, pp.346-371.

community, of the State, of the country, of the group to which one might belong, or at the level of the whole world.

It does not need much argument to recognise that, for individual development, an understanding of the environment and its challenges and of the rules that govern natural phenomena or social behaviour, the ability to evaluate and make judgments that would stand the test of the time, and the ability to receive and assimilate information and convert it into knowledge and then into wisdom, are important. Education, the environment at home, social interactions and experience, and logical reasoning, play an important role in acquisition of the above-mentioned understanding and abilities. The scientific method plays a crucial role in this process by providing a framework for understanding the environment and its challenges and the rules of natural and social phenomena, for evaluating and arriving at opinions and decisions that would stand the test of time, and for logical reasoning.

For collective development, one needs to delineate precisely the goals—both short-term and long-term. The scientific method is an important tool for assessment of the merit of such goals and for ensuring that they are achievable within the means available. This is crucial for defining strategies for using our assets and containing our liabilities to the maximum possible extent, in real time and on a continuing basis, for achieving the desired goals.

History of the last 500 years or so clearly shows that whichever country or society followed the scientific approach—knowingly or unknowingly—succeeded, and whichever country didn't, had to pay a heavy penalty. In our own case, we did succeed after 1947 but not as much as we should have. The basis of our success has been our commitment to science and the scientific method as stated in the Scientific Policy Resolution of 1958, and as has been clear from the policies of our Prime Ministers such as

Jawaharlal Nehru and Indira Gandhi. The reason for our failures has been that we have not been committed enough to the scientific method, outlook or approach. Even our planning process has been devoid of such an approach. This point has been discussed in more detail in the Statement on Scientific Temper that was brought out in 1981 by a group of leading intellectuals of the country, and published by Nehru Centre, Mumbai (see Chapter XVI of the present book).

Much has been written about the development of science and technology in ancient, medieval, and modern India. In view of the above-mentioned relationship between scientific temper and development, and the well-established role of science and technology in development, it would be relevant to ask as to what the extent was to which scientific temper existed, developed and was used, consciously or subconsciously, in ancient and medieval India, say, till the end of the last century.

We shall first state what scientific temper is and then go on to talk briefly about the scientific method and the attributes of knowledge obtained through the application of this method. We shall then discuss the extent to which scientific method was practised, and the attributes of knowledge gained through the application of the scientific method recognised, during our scientific history with special reference to biology.

II. WHAT IS SCIENTIFIC TEMPER?

In our perception, the scientific temper implies the following:

(a) An understanding of the scientific method.
(b) An understanding of the attributes of knowledge gained by the application of the scientific method.

(c) Recognition that this knowledge would be the closest approximation to truth at a given time.
(d) Rejection of what is totally incompatible with such knowledge and cannot be indisputably established by using the scientific approach or method.
(e) Using the scientific method or approach in our everyday life.

III. THE SCIENTIFIC METHOD

Science is far more than just physics, chemistry, biology, astronomy and mathematics, or increasing agricultural or industrial production. It is a question of ideas and a way of thinking; it is a culture and a philosophy of life, a philosophy which allows us to pursue truth without any prejudgment. What, then, is this attitude of mind, this culture and this philosophy of life? It turns out that all these—and other—concomitants and attributes of science emanate primarily from the method that science uses to acquire knowledge.

There are four distinct steps in the method of science—the framing of the question, the framing of a hypothesis, the doing of an experiment, and arriving at the answer which may be a fact or a generalization in the form of a theory or law. At each step in this sequence and in going from one step to the next, we use existing knowledge and logical reasoning.

The *question* in science arises out of careful observation or careful analysis of existing knowledge; there is no third origin of a question! In fact, if you have framed your question properly, you are already on your way to finding the answer. What, then, is a well-framed question? A well-framed question is the one for the answer to which means are available within the framework of the method of science.

Such a question should lead to a hypothesis which can then be tested by an experiment. Let us take an example.

Out of the million animal species that inhabit our earth, nearly seven hundred thousand, i.e., some 70 per cent of them (all insects), have six legs—indeed, a remarkable observation, having many implications. Well-framed questions arising out of this observation would be: How did all the seven hundred thousand or so species come to have six legs? Do these species have other common features? Could they have originated in nature from a common ancestor? These are well-framed questions. The more carefully we observe, the better framed our question would be.

The second step, the *hypothesis,* is an answer we may consider possible. The single most important attribute of a good hypothesis—a scientific hypothesis—is that it must be testable. A testable hypothesis is one which can be tested by an experiment or on the basis of which a testable prediction can be made. For example, a commonplace observation is that objects, if left unsupported in space on earth, will fall to the ground. One can make many untestable hypotheses to explain this phenomenon. One such hypothesis would be: they fall to the ground because a particular friend of yours wills them to do so. Another would be: they fall to the ground because God desires it that way. None of these hypotheses is testable. Therefore, they are not scientific. On the other hand, if you make the hypothesis that an object unsupported in space on earth falls to the ground because there exists a force of attraction between the object and the earth, you can test the hypothesis.

The third step in the method of science is the *experiment.* The experiment must have one of the following two objectives: either to find an answer to a question, or to prove or disprove a hypothesis. An experiment which does not attempt to do one of these is unlikely to be a good experiment. To do an experiment, one must make an inventory of all the steps in the experiment, collect all the material one needs, and carry out all the steps with the utmost possible care. *More* important, one must record all the observations made immediately and meticulously, paying attention to the smallest detail. And *most* important, one must record everything—expected or unexpected—whether one wished it to be so or not to be so. It is in the doing of an experiment in this manner that the values of objectivity, lack of bias and the exercise of care in what one does, come into existence in science. An unexpected observation could lead to an important discovery. We may recall that penicillin was discovered by Alexander Fleming in this way—because of an entirely unexpected observation he made while trying to test quite a different hypothesis.

The last step in the method of science is the *answer.* The answer generally takes one of two possible forms. It could either be a scientific fact of limited applicability, or a generalisation of wide applicability. An example of the former (a scientific fact) would be the answer to the question "What is the density of the paper this is printed on, or what is its chemical composition?". An example of the latter (a generalisation) would be the theory of relativity or the Mendel's laws of genetics. The use of scientific method generates information which at that time would be the closest approximation to truth, provided the method has been used appropriately and has taken into account the entire body of knowledge that is available at that time.

In going from one step to the other in the scientific method, one uses logical reasoning and brings to command all of one's past experience and knowledge. In fact, knowledge is a consequence of collation of bits of information obtained through the use of the scientific method. Many tools are used in this collation; some of these tools are: drawing of inferences using logical reasoning, classification, analysis, interpolation and extrapolation.

We must also recognise that knowledge—which can be generally defined reasonably precisely if it has been acquired by using the scientific method—is not equivalent to wisdom which is far more difficult to define and would seem to consist of many more intangibles than knowledge. Experience and logical reasoning allow one to convert knowledge into wisdom which leads one to take decisions that would stand the test of time, often on the basis of knowledge or information that is not total or complete. Thus, the process of converting knowledge into wisdom involves interpolation, extrapolation, imagination, analysis of linkages and the ability to take a global view (this list is surely not complete!).

IV. ATTRIBUTES OF KNOWLEDGE GAINED THROUGH THE APPLICATION OF THE SCIENTIFIC METHOD

The Right to Question

All knowledge arrived at by the use of the scientific method is a consequence of questioning. One of the most important attributes of science is, therefore, the right to question. There are no high-priests in science who cannot be questioned. Knowledge advances and science progresses because people exercise their right to question. However, to question existing knowledge (a fact, theory or law) without any rational basis or reason is as unscientific as never to question at all. The reasons for questioning may be a flaw found in an earlier experiment; a known observation which was earlier ignored and which can be shown to be incompatible with the earlier fact, theory or law; an alternative explanation found for the evidence on which the earlier fact, theory or law was based; or new evidence which is incompatible with the existing fact, theory or law. Therefore, science puts a constraint on the freedom to question while making the right to question a 'fundamental' right.

There are No Know-alls in Science

Scientific knowledge never claims to be complete—that is, science does not claim to provide immediate answers to every legitimate question that can be asked at any given time. A scientist can, therefore, say without any feeling of guilt or shame, "I do not know".

Scientific Truths are Truths by Consensus

Knowledge acquired through the application of the method of science represents truths arrived at by consensus based on the method of science. The consensus has to be reached among people of various strata and levels, as long as all of them are knowledgeable in the area concerned and have formed their opinion by using the method of science.

How Does Science Progress?

In science, at a given time, we accept a fact or a theory or a law when:

(1) all the observations made and experiments done until that time support the fact or theory;
(2) the new theory satisfactorily explains all that was explained by the old theory;
(3) no new experiments can be conceived at that time the results of which may not support the theory; and
(4) all predictions made on the basis of the theory upto that time have turned out to be right.

Then, and then alone, is a theory in science accepted.

An existing fact or theory gives way to a new fact or theory when the following criteria are met:

(1) New experimental evidence is obtained which is not in conformity with the existing fact or theory.
(2) The new theory satisfactorily explains all that was explained by the earlier theory plus some additional observations not explained by the earlier theory.
(3) Predictions can be made on the basis of the new fact or theory which could not have been made on the basis of the earlier fact or theory.
(4) Some of these predictions have been tested and every tested prediction has turned out to be right.

In fact, a new discovery in science often disproves an earlier scientific belief, that is, a part or whole of an existing scientific fact, theory or law.

Science Versus Supernatural

Scientific knowledge contradicts the existence of the supernatural and of miracles that defy all of science. Science, therefore, does not seek an explanation of the unknown in terms of another unknown. When a scientist does not know the answer to a question, he says, 'I do not know'. He does not accept an 'unscientific' answer, that is, an answer which is incompatible with the method of science. He tries to find out if a scientific answer already exists. If no such answer exists, he uses the method of science to obtain an answer. His basic premise always is that, if the question is unanswerable, the answer can be found only through the method of science.

Predictability

A scientific fact, theory or law allows one to make testable predictions. For example, the Russian chemist, Mendeleev, predicted in 1869 the existence of several elements such as gallium, scandium and germanium, long before they were discovered, and assigned to them their right places in his periodic table. Darwin and other evolutionary biologists predicted the existence of certain species, such as *Latimeria* (a fish), *Pithecanthropus erectus* (the 'erect ape man') and *Oreophithecus* (another ancestor of man), much before their discovery. In physics, the existence of the fundamental particles, omega minus and neutrino, was predicted beforehand. In astronomy, the planets Neptune and Pluto would probably never have been discovered if astronomers had not *looked for them* following the prediction of their existence.

Scientific Observations are Verifiable and Repeatable

They do not depend on the whims and fancies of individuals. Thus, a time reaction set for 19 ± 1 seconds, will take 19 ± 1 seconds—no more and no less—irrespective of: (i) who carries out the reaction (a child, a man or a woman: Indian, Chinese, African or European); (ii) what you may personally like to happen; (iii) the place where the experiment is done; (iv) the time of the day, month or year when the experiment is done.

Science is Truly International

There is one, and only one, science. Scientists all over the world use the same method (the method of science), employ the same techniques, use the same materials, publish in the same journals, are increasingly beginning to use the same language (that is, English), and form a truly international community in which the professional links are at least as strong, if not stronger, than other links.

Science Refutes Religious Dogma that Distinguishes One Religion from the Other

We have argued elsewhere in detail that science and religious dogma are incompatible (P.M. Bhargava, 'Does science refute religion', *Society and Science*, 1981, Vol. 4, pp 42-50; reproduced as Chapter XI in this book).

V. THE SCIENTIFIC METHOD IN OUR SCIENCE HISTORY

In the discussion that now follows, we would give examples, mostly from biology with which we are more familiar. (It would be interesting to determine whether or not what we say here about biology would also be applicable to other sciences, specially astronomy and chemistry that developed substantially in ancient and medieval India.)

In the scientific method, the question that the individual wishes to answer, arises out of careful observation or careful analysis of existing data. We have had a superb tradition of making careful observations and recording them accurately: our ancestors were indeed compulsive observers. For example, detailed knowledge of various internal and external organs of the body and its various systems was acquired during the Vedic period. In *Atharvaveda,* there are references even to Fallopian tubes and to the relationship between testicles and semen. We were aware of not only bones but also cartilages and ligaments. In the *Charaka Samhita* (believed to have been put together somewhere between 4th century BC and 4th century AD, probably around 100 AD) the total number of bones in the human body has been stated to be 360. As we know the total number of bones in the human body to be 206 today, the chances are that all of them had been identified by Charaka's time.

Susruta's description, even before Charaka, of the anatomy of the human body, within the limitations of the human eye, is breathtakingly comprehensive and analytical. The basic difference between vertebrates and invertebrates was clearly recognised: "some beings stand mainly with the support of skeleton and others with muscles".

Parasara (1st century BC to 1st century AD) gave the details of the internal structure of leaf. His description refers to innumerable small compartments, cell sap and possibly cell wall.

The *Brhataranyakopanishad* (1000-600 BC) compared the human being with a tree as follows: "A man is indeed a mighty tree; his hairs are its leaves and his skin is its outer bark. The blood flows (from the skin) of the man, so does the sap (from the skin) of the tree. Thus blood flows from a wounded man in the same manner as the sap from a tree when it is chopped. Flesh within corresponds to the inner bark; his nerves are as tough as the inner fibres of the tree; his bones lie behind his flesh as the wood lies behind the soft tissue. The marrow of the human bone resembles the pith of the tree". Surely, there is an element of both realism and poetry in these analogies! Susruta, too, gave a more or less detailed account of different parts of a plant, with a tendency to compare the plant parts with those of the human body.

Classification of plants and animals into manageable categories came naturally to our ancestors. Some 740 plants and over 250 animals seem to be referred to in our ancient literature; they were classified in many different ways, for example, on the basis of their medicinal property, domestic utility, or morphological features. The first attempt to classify animals in some rational way is found in the *Chandogya Upanishad,* where classification was based on their mode of origin and development. In this classification, there was a group comprising of organisms that were born out of "heat and moisture of the earth", such as stinging gnats, mosquitoes, lice, flies and bugs. It is interesting that all animals that were small and apparently caused some damage or discomfort, were thought to arise spontaneously, out of the scum of the earth! It was only in the later half of the last century that the theory of spontaneous generation was finally buried by Louis Pasteur, so our ancestors did not do badly at all!

The most elaborate classification of plants was by Parasara who based it largely on morphological considerations such as floral characteristics. He classified plants into families some of which clearly represent families of today, for example, Leguminosae, Crustacea, Cruciferae, Cucurbitacea, Kapucynacea and Compositae. The tragedy is that such classification was not improved upon subsequently.

However, the observations our ancestors of the ancient period made, rarely seem to have led to precise questions which would allow the framing of a hypothesis and testing of the hypothesis through means such as an experiment. This stands out in contrast to the development of what is today called "Western science" in which framing of a question was extremely important. Thus, Newton wouldn't have discovered his laws of gravity if he hadn't asked the question as to why objects left unsupported in space fall to the ground—question based on common-place observations.

In today's science, often, an important question relates to the resolution of an apparent paradox. It would be interesting to look at the history of ancient and medieval Indian biology from this point of view—that is, to determine if questions were asked specifically to resolve certain paradoxes. We have not made such an attempt yet.

As we have already mentioned, questions in science also arise from analysis of existing information. Such analysis often implies collation of information from various sources in different areas obtained by different people at different times. It was such an analysis that led to the discovery, for example, of reverse transcriptase (an enzyme that is contained in certain viruses and allows the synthesis of DNA from RNA; a Nobel prize was awarded for the discovery of this enzyme which has played a crucial role in genetic engineering). Similarly, our current views on the predictability of monsoon taking into account as distant an element as the El-Nino current of South America, depend on a similar analysis of data from a vast variety of sources. In our ancient and medieval biology, we do not seem to have many (any?) cases of an interesting question being asked following analysis of existing data or information. In fact, in our opinion, if this capacity had existed in our ancestors, the course of development of biology in India would have been different. Even though Parasara classified plants into families, some of which clearly represent families of today, the relationship between various classes was not analysed. If this had been done, a more systematic classification would have almost certainly emerged many centuries ahead of Linnaeus. This analysis would have required recognition of the importance of dealing with many parameters at the time of classification.

Coming to the second step in the scientific method, the framing of hypothesis, the history of our ancient and medieval biology is replete with hypotheses. Unfortunately, these hypotheses did not have a base in a specific question which, in turn, would have been based on careful observation or analysis. These hypotheses were essentially statements, very often without any scientific basis; they could not be considered as axioms because they were not self-evident. These hypotheses, even though not having any real or solid foundation from the point of view of today's science, nevertheless seemed to have served as a base for putting up large superstructures on them where, again, each statement had to be accepted as such, without questioning. The entire theory of Ayurveda is based on such statements, hypotheses or superstructures.

We give below examples of some such hypotheses from our ancient biological literature. An important criterion of a scientific hypothesis is that it should be possible to design experiments that would disprove it just as it should be possible to design experiments that would prove it. These criteria are not satisfied by any of the following statements which are no more than bad hypotheses for they have not been proven to be true.

(a) Susruta in his treatise has given a remarkable description of the foetus at various stages of development. He has also prescribed the care that needs to be taken during pregnancy. All this has been found to be fairly accurate. Yet, the same Susruta attributes the reasons for congenital defects that, for example, dwarfs and hunch-backs suffer from, to the fact that the mother's desires during pregnancy were left ungratified or repressed. There were several other speculations about cause and effect during pregnancy that we know today are not true. Both Charaka and Susruta also believed that the fertilised ovum (or the foetus) developed by palingenesis and not by epigenesis. In other words, all the organs were potentially present in miniature form in the fertilised ovum or the seed, and unfolded in a certain order during the growth of the foetus. Even Shankara shared this view. We *now* know that this is not true.

(b) Charaka and Susruta believed that diseases were caused by a disturbance in the equilibrium of the three Ayurvedic humours, and that this disturbance was often a *direct* cause of the disease. They also recognised several remote causes, both external and internal, for example, entry of toxic materials from outside, errors of living, natural decay from old age, and climate or weather, that could play a role in the manifestation of disease. While they were right about the remote causes, their above-mentioned hypothesis about the direct causes of disease is totally untenable.

(c) There is a mention in *Mahabharata* that plants are sensitive to heat, cold, sound of thunder, and odours, and experience pleasure and pain. While one can substantiate the fact that plants are temperature-sensitive, and one could perhaps stretch one's imagination to include odours by thinking of the possibility that poisonous gases could have a deleterious effect on plants, current scientific evidence does not substantiate the suggested sensitivity of plants to the sound of thunder (or to music to which claims were made subsequently), or their capacity to experience pleasure or pain as we understand these feelings.

(d) It is stated that "children born with the 'best factors' available are destined to be handsome, virtuous, long-lived, generous, beautiful and responsible in their conduct". While this hypothesis does imply recognition of hereditary factors, it hardly takes into account the influence of environmental factors in the manifestation of genetic abilities.

(e) In *Agnipurana,* it is stated that the sex of the child is determined by the position of the foetus; this is not true. Susruta has given a remarkably accurate description of the safe and the unsafe periods in a woman's menstrual cycle, but he has also gone on to make the erroneous hypothesis that the child is male if conceived on even days and female if conceived on odd days beginning from the day of the cycle. He also erroneously says that sex is determined by the strength of the sperm or the egg; thus the progeny is male if the sperm is stronger and female if the egg is stronger; when the strength of both the sperm and the egg match, the offspring is supposed to be a hermaphrodite.

Coming to experiments, the only area where there seems to be substantial evidence of experimentation is surgery, and that is, perhaps, the reason why surgery has been the most illustrious branch of Indian medicine—perhaps, one of the most illustrious in all of ancient or mediaeval Indian science. Thus, fifteen different methods were described by Susruta for the extraction of a foreign body loosely or firmly embedded in the tissues, a magnet being used for iron particles. And, Indian doctors in the ancient period achieved such perfection in plastic surgery that European scientists of the 19th century borrowed several methods from them. Susruta also discovered the art of cataract-crouching which was unknown to ancient Greece and Egypt. Limbs were amputated, abdominal operations performed, fractures set, dislocations, hernia and ruptures reduced and haemorrhoids removed—all with an amazing rate of success. The earliest of rhinoplasties appear to have been performed in India in 1600 BC; subsequently, the technique of rhinoplasty operation spread from India through Arabia and Persia, to Egypt and from there to Italy.

Indeed, the basis of these successes was the fact that dissection of dead bodies was considered indispensable for a successful student of surgery and that such bodies were made available for various experiments. It would be interesting to analyse the reasons as to why the experimental method that succeeded so well in the development of surgery in ancient India, was not used in other areas of biology with the same rigour.

We have mentioned that the use of scientific method often generates information, and to convert information into knowledge one resorts to various approaches such as collation of information, drawing of inferences, trial and error, classification, etc. Our ancestors certainly knew how to convert information into knowledge by using all these techniques. We must, however, immediately add that none of the techniques that we have mentioned for conversion of information into knowledge, is fool-proof. The validity of the knowledge obtained by these techniques has, therefore, to be continuously checked and rechecked by further experimentation and questioning as new techniques and new evidence become available. This is unfortunately where we failed. Thus, we have instances of important conclusions arrived at by using the above techniques that have stood the test of time, just as we have instances to the contrary. We would now like to give examples of both.

VI. EXAMPLES WHERE CORRECT CONCLUSIONS WERE ARRIVED AT THROUGH COLLATION OF INFORMATION, DRAWING OF INFERENCES, CLASSIFICATION OF INFORMATION, ANALYSIS, EXTRAPOLATION, INTRAPOLATION, AND TRIAL AND ERROR

Our ancestors had rightly concluded that life, including all biological processes within us are dependent on the generation of heat. Our ancient medieval literature states: "This body heat comes out of food which also nourishes and maintains the organism through its metabolic transformations. Ingested food and drink pass into the stomach and become minutely dispersed by the digestive fluid present there; their assimilable contents then turn into a sweet, frothy, mucus-like fluid. This process of digestion, carried out by *agni* (digestive fire), continues until the fluid becomes acid, issues out of stomach and excites the secretion of thin bile. At this stage, it is an assimilable, nutritive fluid known as *rasa,* which is pumped by the heart through twenty-four major channels and permeates the entire system. *Rasa* constantly moistens, nourishes, maintains and irrigates the organism by processes which are not completely understood." Mostly today's state-of-the-art, and incorrect only in detail!

The conclusions arrived at in regard to the reproductive process, both in animals and in plants, in ancient India were truly impressive. Although there does not appear to be enough evidence of the knowledge of sexuality in plants in the Harappan culture, sexual reproduction in higher plants as well as in higher animals is mentioned in the pre-Buddhistic *Kathopanishad* as being similar. Seeds and flowers were believed to be produced by the cooperation or union of different sexes. Pollen was believed by Amara to be analogous to the female menstrual fluid. In the *Brahmanas*—a constituent of the *Vedas*—there are many references to conception and to child-birth. As has been already mentioned, the testicles were recognised as being responsible for the production of semen. It was also recognised that the semen should get amalgamated with the contribution of the woman in her womb; unless this happened, pregnancy could not be established.

Garbhopanishad gives a detailed and fascinating description of the day-to-day and monthly development of the human embryo through its various stages, from conception to delivery. Susruta gave the best time for conception from the fourth to the twelfth day from the date of the beginning of the menstrual flow: precisely what is recommended for a 22-day menstrual cycle today! Imagine the amount of information he must have collected to arrive at this conclusion, and that too how, and in what kind of a culture! (May be, our today's perceptions of that culture are inadequate.)

The role of the umbilical cord and of the navel was amazingly well recognised. "The *dhamanis* in the foetus take their rise from the umbilical cord, thus bringing nourishment from the mother", and the navel in the foetus was rightly stated to be the source and origin of the entire vascular system. To continue the quotation, "The embryo is held at the navel. It grows without taking food, that is, there is no effort made on the part of the embryo to take food and no food is specially served to it. The food in its final form, is assimilated automatically and directly into the system of the embryo. The child is nourished of its own accord as it were. The mother is not conscious of the nourishment given to the young one below her heart." Could it have been said better?

The animal body was recognised to be sustained and nourished by blood which was "conveyed through a large number of channels to every part of the body". Existence of capillaries was recognised in numbers that were impossible to count. It was stated that urine is formed by draining of the waste or refuse matter in the body by water. The water content of the urine was correctly concluded as derived from the drinking water and from the moisture of the food taken in. Urine was, therefore, correctly thought of as a body fluid which served to eliminate waste metabolic products not required by our body.

The growth of a plant was recognised to depend on soil, water and season. It was recognised that light had something to do with the process of manufacture of food by plants and storage of energy in their body.

Charaka made another highly perceptive and logical statement when he said that diagnosis of a disease should depend on (i) theoretical knowledge of the possible causes and symptoms of diseases, (ii) meticulous observation of the patient's symptoms and complaints, and (ill) inferences based on previous experience.

One of the most remarkable deductions made in the history of Indian medicine was in regard to small pox. In fact, the impression that all of India was in a state of rapid decline in the late 18th century, is certainly argued against by the fact that inoculation against small pox was practised in the subcontinent at this time, and long before it became generally acceptable in Europe. It was unknown in Europe till 1720, when the wife of the then British Ambassador in Turkey, having got her children successfully inoculated, advocated its introduction into Britain.

The farmers of the Vedic period were aware of the possibility of improving the fertility of the soil by rotation of crops—a concept that developed in the West very much later. Rice was grown in summer and pulses in winter. References to rotation appear in *Rigveda* and *Yajurveda.* Thus rye-grass and cloves were grown with wheat, barley or oats, and beans with peas.

The ancient cultivators knew how to select the seeds

and what to sow when and where; they recognised the need of replenishing the nutrients of the soil by manures. The later Vedic agricultural farmers seemed to be fully conversant with the use of organic matter such as appropriately processed cow dung, bones, blood, and plant products such as the straws of barley. These manures are today known to contain nitrogen, phosphorous and potassium.

From 300 to 200 B.C. onwards, the early stages of the germination of seeds and the factors governing germination (such as proper season, good soil, water, vitality of the seeds, and proper care) were clearly recognised. Kautilya's *Arthasastra* mentions the effect of temperature on germination; it gave specific conditions required for germination of different kinds of seeds.

We had learned how to determine the age of animals from sequential changes in their teeth, and we knew how to train animals and to exercise control over them. Cattle breeding appears to have been one of the important aspects of animal husbandry practice in ancient India.

VII. EXAMPLES OF INCORRECT CONCLUSIONS ARRIVED AT BY USING THE SAME TECHNIQUES AS MENTIONED ABOVE

i. According to Susruta, the intervention of a superior agent was absolutely essential for the origin of life on our planet. We today understand the basic sequence of events that led to the origin of life on our planet some 12 to 15 billion years ago/and in that sequence there is no place for any superior agent.

ii. In the Upanishads, evolution on earth is explained as follows: ether sprang from earth, air from earth, fire from air, water from fire, earth from water, herbs from earth, food from herbs, seed from food, and men from seed.

This, of course, is a far cry from what we know today about evolution.

iii. Our ancient postulate that five physical elements—ether, wind, water, fire and earth—constitute the human body is, of course, totally untenable.

iv. The theory of spontaneous generation of lower forms of life such as maggots and worms, is deeply rooted in our ancient culture. Pasteur disproved this theory dramatically in the last century.

v. The food that we eat was stated to contain five classes of nutrients: the earth-compounds, the aqueous-compounds, the tejas-compounds, the vayu-compounds and the akasa-compounds. The earth-compounds were postulated to form the hard matter of the body, the tejas-compounds to provide the metabolic heat, the vayu-compounds to serve as sources of the motor-force in the organism, the aqueous-compounds to furnish the watery parts of the body, and the akasa-compounds to contribute to the fine etheric essence that was considered to be the vehicle of conscious life. These concepts do not relate at all to our knowledge of nutrition as of today.

vi. In the *Brahmanas,* Charaka is quoted as saying that the offspring derives its softer tissues like skin, flesh, navel, and intestines from the mother, and the harder tissues such as bones, blood vessels, veins, etc., from the father. This erroneous belief, perhaps, satisfied the male ego!

vii. It was believed that mother's nutrition could determine the sex of the child. For a male child, ghee and milk were recommended, and for a female child, oil and beans. Another example of male chauvinism!

viii. In the *Agnipurana,* heart was considered to be the seat of consciousness and the centre of our nervous system. It was only in the Tantric writings between the 8th and the 14th century AD that the seat of consciousness was transferred from the heart to the brain. It was even proposed that one pair of *dhamanis* going from the heart to the head was engaged in conducting sensory currents pertaining to sound, smell and taste. Perhaps, this was so because all the sensory perception was lost with the stopping of the beating of the heart. Today, we know that in a clinically alive person, the heart could still be beating, inspite of all the sensory perceptions being lost. Susruta also stated the precise location of the soul in the body! The soul or *jiva* was supposed to reside in the upper cerebrum but could traverse the whole cerebrospinal axis up and down. There is no place in modern biology for such a concept of "soul".

ix. The Indian traditional medicine contained in the *Ayurveda,* arose from the notion that a body remained healthy if there was equilibrium between the three humours, *vayu, pitta* and *kapha,* present in our body. These three humours were stated to govern and activate the entire gamut of biological processes from conception onwards. The description and characteristics of these three humours in *Susrutha Samhita* does not even remotely correspond to our today's knowledge either about the body or about its functions.

x. The beliefs that were codified in our ancient writing, regarding the position of a pimple determining the immediate fate of an individual, make amusing reading. Thus, "pimples appearing on the hands, fingers and belly, lead to the acquisition of wealth, fortune and grief, respectively; on the navel, to finding food and drink; those beneath the navel, to loss of wealth through theft; on the pelvis, to wealth, on the thighs, to the procurement of a vehicle and a wife; on the knees, to loss on account of enemies; and on the ankles, to trouble while travelling and during confinement".

xi. It seems likely that Manu's belief in the caste system that led to his systematising the entire system that has been at the base of many of our present-day problems, was deeply rooted in his erroneous perception of the hereditary transmissibility of acquired characters.

xii. Our ancient literature is replete with examples of scientifically untenable male chauvinistic statements. Here are translations of two such *slokas* (note the part in italics):

"When (at the time of coitus) the blood (of the woman) exceeds the sperm (of man), a female will be born; when the sperm exceeds the blood, a male; when both are equal, a hermaphrodite. *Hence, one ought to take tonics that increase one's sperm.*"

"A man ought to have sexual union with his wife when the Kendlas and the *Trikona* houses are occupied by benefics, when the Moon and the *Lagna* are conjoined with benefics, when malefics are posited in the 3rd, 6th and 11th houses and when there are planetary combinations *ensuring the birth of a male.*"

VIII. DID OUR SCIENTISTS OF YORE RECOGNISE THE ATTRIBUTES OF SCIENTIFIC KNOWLEDGE THAT SCIENTIFIC TEMPER DEMANDS?

Everyone Has the Right to Question

Lord Buddha who lived in the 4th-3rd century BC said:

Believe nothing
Merely because you have been told it
Or because it is traditional
Or because you yourself imagined it
Do not believe what your teacher tells you
Merely out of respect for the teacher
But whatever, after due examination and analysis
You find to be conducive to the good
The benefit
The welfare of all beings
That doctrine believe and cling to
And take it as your guide ...

Yet, the culture of questioning following the norms demanded by the scientific method, never really took root in India in the ancient or the medieval period. (Even today, scientific questioning, although far more common than, say, even 50 years ago, has not become a part of our culture and thought process.) One important reason was blind respect for age, position, authority and power, which respect became a part of our cultural, social and religious traditions. One could ask a question to seek a clarification from one's elders or teachers, or those who were in power or had authority, but one did not question the validity of their statements. In a hierarchical society that ours has always been—and continues to be—there were no two people who could establish an equal relationship with another. Further, the knowledge system in science at least, was based on the concept of "high priests" who were not to be questioned. This was one of the main reasons for the stagnation of our knowledge and our reluctance to imbibe the advances that occurred elsewhere—specially from the 15th century onwards in Europe. Consequently, for example, the Ayurvedic practice of today is based virtually entirely on what was written in ancient texts by the high priests of that time. The concept of high priests who may not be questioned is so deeply ingrained in our culture, with its origin during the Vedic period, that even our present-day society is replete with Godmen who have an immense following, so much so that—may be—one out of every three adults in the country may be today a follower of one or the other Godman.

Scientific Knowledge is Never Complete and There are No "Know-alls" in Science

One of the most significant deficiencies of our ancient science has been that it has purported to be complete with answers provided to all possible questions. It is for this reason that many in the country believe even today that our ancient literature contains answers to every question that one may ask today, and that everything that is known today—from aeroplanes to nuclear devices—was known to our ancestors. Ayurveda is thus presented as a complete medical system with a cure for virtually everything. This view is supported by the fact that, barring minor nuances of interpretation, basically the Ayurvedic repertoire of today—both in regard to knowledge and practice—is probably not very different from what it was more than 2000 years ago. This stands out in contradiction to one of the important attributes of scientific knowledge—that it progresses but is never complete.

Scientific Truths are Truths by Consensus

A large number of individuals take part—individually and collectively—in the process of discovering scientific truths, their consolidation and subsequent acceptance. The consensus is arrived at amongst equals with each one having the right to verify and challenge everyone else. Of course, everyone who takes part in this process and whose opinion counts, has to be appropriately trained and committed to the use of the scientific method and acceptance of the attributes of knowledge gained through this method. It is obvious that for arriving at a consensus that would be likely to stand the test of time, there has to be much discussion, debate and questioning. The *gurukul* tradition and the *guru-shishya* relationship that was at the base of transmission of knowledge in our ancient culture, did not have the ambience to create equals who could question each other and carry out a discussion of the kind mentioned above. Consequently, our ancient and medieval science was not a result of consensus but a collection of pronouncements made, no doubt, by highly knowledgeable people. Truths do not appear to have been established by verification, discussion and doubting, and the resolution of doubts, by equals.

Science Progresses by Disproving and this Process Follows Certain Rules

It is, indeed, the fact that science progresses by disproving that makes it evolutionary. All the evidence that we have been able to gather in the area of biology suggests that the method of disproving was not used to advance scientific knowledge. In our tradition, if anyone had a new idea, he propagated it for others to accept or reject, without his disproving what was believed to be the truth earlier. We have not come across any instance in which a later scientist had openly criticised someone who came earlier and actually proven him wrong so that the earlier belief was buried forever—just as Dalton's atomic theory is buried forever.

Science Contradicts the Existence of the Supernatural

Belief in the supernatural and related magico-religious practices, was central to all science of the past: from astrology to Ayurveda. For example, the belief in sins, demons and black magic as the cause of disease was deep-rooted and widespread in ancient India. It appears that during the Atharvavedic period, there existed two main types of healing arts. The first type depended largely upon incantation of magical verses and sacrificial practices to bring about cures. The second type, while also using magical formulae, relied basically on the empirical or rational use of herbs and other medicaments. Thus, herbs were used in combination with spells, and diseases were sought to be cured by propitiating the demons: cures that had the authority of *Upanishads* and the *Sutras* (800-300 BC) behind them. There were specific mantras for particular diseases. Thus, there are mantras addressed to *Surya* to cure heart disease and jaundice. Similarly, there is a mantra to increase the power of sight and to cure certain diseases of the eye. One would not be exaggerating if one says that this is unmitigated nonsense. If the magico-religious elements had been shed, and our ancient system of medicine had confined itself to adhering to and relying upon empirical observations which could then be modified as the database increased, perhaps the evolution of medicine in India and, consequently, around the world, might have taken a different course.

The supremacy of magico-religious medicine during the Atharvavedic period is more than evident in their belief in the wonders of an amulet. "An amulet was looked upon as a weapon, an instrument, which protects the wearer against misfortune and disease." And there are many hymns in the *Atharvaveda* to be recited at the time of binding an amulet". The material in the amulet depended on the effect that was desired—from curing hereditary diseases to obtaining a male child!

Diseases were believed to be caused by a variety of unverifiable and ill-defined external agents and actions—such as wrath of Gods, possession by demons and evil spirits, sorcery and the like—which, by definition could never be shown to exist or experimented upon. As time passed, the evil influence of planets and stars also came to be regarded as a cause of disease. And it wasn't all that crude either: *specific* diseases were believed to be caused by *specific* demons or spirits. We, of course, do not believe that these ideas can be equated with the causation of different diseases by *specific* microorganisms that we know of today!

Science Allows One to Make Testable Predictions

This attribute of science was unquestionably recognised by our scientists of the past though there is little evidence of this attribute being used to actually advance the frontiers of biology. On the other hand, the predictive value of Indian astronomy seems to have been its bread and butter and is what gave it a sound footing.

What would be worth investigating in some detail *is* whether the considerable success that our astronomers had with their predictions was partly responsibly for laying the foundations of the unscientific system of astrology.

Scientific Observations are Verifiable and Repeatable

We have no doubt that many of the observations made by our ancestors *that have stood the test of time* and are compatible with today's science, must have been verified by subsequent observers in history. Similarly, many conclusions arrived at by the limited experimentation to which our ancestors resorted to as in surgery, or by calculation as in astronomy, must have been verified through subsequent repetition. Therefore, in a way this criterion of scientific knowledge was probably satisfied by at least a part of the knowledge gathered at that time—the part that is valid as of today. However, the important question we must ask is: whether or not verification and repetition with an open and unbiased mind was a part of the methodology of authentification of the scientific knowledge in the past. This would merit a detailed study which we have not yet done. We suspect that this was so but only on a very limited scale. We know that a substantial body of knowledge gathered in the past—specially that which was based on conclusions arrived at on the basis of inadequate observation or experimentation—has not been found to be verifiable and repeatable today. It must have been found to be so by others as well in our past but, perhaps, due to the prevailing social milieu, the person who could not verify or repeat what had been said earlier, thought it wise not to document *his* failure. He may have merely documented his new idea without reference to the past. We believe there is a strong case for a systematic study of original source material that is available to us today to determine to what extent a reference has been made to all the earlier work and in what context. (Perhaps, such a study has not been done so far.)

Science is International

What we should perhaps say is that *valid* science is international. In fact, international acceptance is today considered an important criterion of validity of a scientific theory or idea, observation or experiment. Our astronomy, mathematics, technological processes and later practices such as vaccination, did receive this validation; the same was true of surgery. However, Ayurveda *as a system* did not receive this validation, though the efficacy of certain drugs which form a part of the Ayurvedic pharmacopoeia was subsequently confirmed internationally; reserpine would be one example.

International validation is a reciprocal process. We may, therefore, also ask as to what extent did our biologists in particular and scientists in general attempt to validate what had been discovered or claimed elsewhere in the world—at least in geographical areas with which we had regular communication. It is this mutuality of validation that makes one a part of the international structure of science which brings its own rewards. We must also distinguish this process of mutual validation from the indirect influences that one country's science may exercise over another country's science—influences which are a consequence of a large number of factors that operate over a period of time. (There is an intangibility about such "influences" which is not there in the process of scientific validation.) Again, it would be worth examining our ancient and medieval scientific literature and source material to determine to what extent our scientists engaged in validation of knowledge gathered elsewhere, and to what extent they were influenced by it, making a distinction between the two processes.

Science Refutes Religious Dogma

Our science in the ancient and the mediaeval period does not appear to have refuted religious dogma. On the other hand, it seems to have become subservient to it and even sought the seal of approval of dogmatic and other irrational beliefs that were either an integral part of or were deeply intermeshed with religion. Some examples have already been given.

Were There Other Attributes of Scientific Temper in Our Knowledge of the Yore?

We did not recognise the fact that scientific knowledge at any given time is only the closest approximation to truth; we equated it with immutable truth. This tradition continues even today. Therefore, we rejected what was not compatible with it even if it could be independently established by using the scientific approach.

Our daily life was governed by the dictates of religion, custom, convention and tradition, rather than the scientific approach. We accepted science only if it came to us through these other approaches and not as an independent body of verifiable truth.

Conclusion

It seems to us clear that even though Indian science reached remarkable and commendable peaks in certain areas, the scientific method and the scientific temper as a methodology, approach or way of life, were alien to us. This in itself, perhaps, could not be considered as a major disadvantage. What, we believe, has been a disaster is that the scientific method and the scientific temper continued to remain alien to us even when they became a major force in Western thought, at least from the thirteenth century onwards, specially, with the beginning of the Renaissance in Europe with Leonardo da Vinci. And scientific temper and method continue to remain alien to us even today, more than a century after the introduction of Western science in the country—when other countries with equally strong local traditions in science, such as China, Japan and Vietnam, have recognised their validity.

One may ask why the scientific method did not develop and scientific temper did not take root in India in spite of the fact that India has had a five thousand year-old tradition of science, unlike Europe with the exception of Greece. We give below some of the possible reasons:

1. The above-mentioned tradition itself could have acted as a block to the development of scientific temper and the acceptance of the scientific method on account of an in-built resistance to change that has characterized all early societies.

2. In a hierarchical society, specially one in which knowledge is considered as the confine of a select number of people identified by birth, knowledge becomes a tool of perpetuating vested interests. Under such circumstances, development of a culture of questioning which is an integral part of the scientific method, would pose a serious threat to the continuance of these vested interests.

Authors' Note: In preparing this article, we have drawn extensively on our earlier writings in these and related areas, specially the following which provide references to the statements made here:

- *A New Approach Towards the Study and Analysis of the History of Development of Biology in India*, P M Bhargava and Chandana Chakrabarti. Published as a 58-page Occasional Paper of the Project of History of Indian Science, Philosophy and Culture, by Nehru Memorial Museum and Library, New Delhi, 1992. Also published in *Science, Philosophy and Culture in Historical Perspectives*, PHISPC Monograph series on History of Philosophy, Science and Culture in India, Vol.4, Ed. D P Chattopadhyay and Ravinder Kumar, published by Project of History of Indian Science, Philosophy and Culture, Indian Council of Philosophical Research, New Delhi, 1995. (199 references).

3. Finally, the tradition of experimentation was lacking in our ancient and mediaeval culture—even in Gautama's *Nyaya Shastra.* Experimentation to test a hypothesis based on observation or analysis of existing information, is the key to all modem scientific inquiry and progress, and an important step in the scientific method. Wherever we did experiments as in relation to surgery, the knowledge gained has stood the test of time.

- *Some Aspects of the Ancient and Traditional Biology: Looking Through the Magnifying Glass of the Modern*, P M Bhargava and Chandana Chakrabarti, in PRAKRTI, The Integral Vision, Vol.4, *The Nature of Matter*, Ed. Jayant V Narlikar, published by Indira Gandhi National Centre for the Arts and D K Printworld (P) Ltd, New Delhi. (22 references.)

XIX

SECULARISM AND SCIENTIFIC TEMPER*

P M Bhargava **and** *Chandana Chakrabarti*

What is Secularism?

There has been a severe lack of understanding in the country about the meaning of the term, secularism; it has often been taken to imply tolerance of religion other that one's own, and no more. The Longman's dictionary defines secularism as a "system of social teaching or organization which allows no power to religion or the Church". The Oxford dictionary says that being secular means being "skeptical of religious truth or opposed to religious education". Secularism is, therefore, not merely tolerance of other religions while continuing to have faith in the dogma of one's own religion. It has an element of activism rather than representing merely a passive belief. It involves as much denial as acceptance. Secularism does not argue against personal beliefs—as long as such beliefs are not institutionalised, or imposed on anyone else, or lead to a social conflict.

What is Scientific Temper?

There are two ways in which the practice of science and the knowledge generated through it, has increasingly influenced man and society: (1) through providing material gain,

---

* This article appeared in *Secularism in India*, Ed. M.S. Gore, Vindhya Prakasan Pvt Ltd Co., 1991, pp.162-173.

comfort or convenience of one kind or another, or material means of acquiring power or improving the quality of life; and (2) through influencing our thought processes, views and value system, and the manner in which we arrive at decisions or form opinions—including decisions or opinions that are related to the practical application or use of the scientific knowledge. Scientific temper is the quality of mind that defines and determines the latter aspect of relationship between science and society.

Scientific temper was a phrase much in Jawaharlal Nehru's vernacular. He reiterated it not only in speaking of science, but also in exhorting his countrymen in diverse contexts. The phrase is an attractive one and has both brevity and comprehensiveness, for *temper* indicates all the hues of man's thinking, nicely qualified to the plausible and rational with the adjective *scientific*. It implies readiness to consider *all* facts, and not merely facts which are in consonance with one's own thinking or comfort; it obligates one to an active search for such information by study and questioning. It also implies a trust that events are shaped by the fruits of man's labour, and a healthy skepticism towards all claims of supernatural participation in his affairs. In fact, the scientific attitude is simply one of adherence to facts, an ability to revise opinions and a rational skepticism to claims for non-material intervention in our affairs. The spirit of inquiry and the acceptance of the right to question and be questioned, are fundamental to scientific temper. It is, therefore, incompatible with the acceptance of authorities of all kinds, or of 'high priests', who may not be questioned.

Origins of Science and Religion

Since secularism essentially means negation of the dogmatic aspect of religion, the prerequisite for it to prevail is the prevalence of scientific temper which questions dogma. And scientific temper arises out of the practice of science.

Therefore, to further our understanding of the relationship between secularism and scientific temper, it would be useful to trace the origins of science and religion. Both of them were a consequence of the evolution of intelligence in man; the irony, however, is that science and religion, which today stand in direct contrast to each other, arose out of the quest for answers to the *same* questions that bothered primitive man. Man was intrigued with everything that he saw around him and this led him to ask himself questions about the non-living materials he saw around him, such as water, air, earth and minerals; the physical phenomena he witnessed, such as light, heat, sound, thunder and lightning; the extra-terrestrial objects and phenomena he observed, as the periodical rising of the sun, the moon, the stars, the passage of the planets through the various constellations and, of course, the eclipses; and questions about the living things that he saw around him, for example, the recurrent phenomena of birth, death and disease.

The fear and awe that arose from his inability to understand his surroundings, led him to believe in the existence of a supernatural power and to construct self-consistent systems of belief that were to be accepted entirely on faith and without questioning; such systems of belief provided him with plausible answers to his questions. Since the basic premise of these beliefs was an unquestioned acceptance, dogma became an important and indispensable factor. It is this kind of intellectual effort that, perhaps, led to the development of religion, both pagan and codified—the codified religions including Hinduism, Buddhism, Judaism, Christianity and Islam. The main difference between the pagan and codified religions has been in regard to the basic premises: how far one could go before the logic breaks down.

However, as man progressed and the total fund of human knowledge increased, the inconsistency of the dogmatic aspect of religion showed up and man began to question the basic premise of religion itself. He began probing the same questions that he had earlier tried to answer through religion, with renewed effort, from which emerged what we formally know today as the method of science. It soon became apparent that this method could not only be used as a tool which would satisfy human curiosity much more than religion had done so far, but it also opened up new areas for investigation that had so far been hidden or even prohibited.

The phenomenon snow-balled from the thirteenth century onwards, and we had Roger Bacon, Leonardo da Vinci, Copernicus, Francis Bacon, Galileo, Rene Descartes and Issac Newton, amongst others, to give new dimensions to the method of science, that is, the newly developed art of questioning. The answers that emerged did not demand acceptance on the basis of faith alone; moreover, they were testable and verifiable, and did not depend on the whims and fancies—or the likes and dislikes—of an individual or a group of individuals. The explanation provided by science through the use of the method of science, was eventually always found to be more appealing to reason. Science, therefore, grew up, so to say, as a competitor to religion, answering more successfully, the same questions that religion had earlier attempted to answer, and thus coming into direct conflict with religion.

Religion soon became a hindrance to the progress of science, and led to the persecution of scientists. Consequently, history records the conflict between Copernicus and Galileo on the one side and the Church on the other; and between the Church and Charles Darwin's theory of evolution that was so ably extended by Thomas Huxley to the evolution of man.

Copernicus had to recant because he said that it was not the sun that goes round our Earth, but Earth that goes round the sun. Galileo, a follower of Copernicus, died under house arrest imposed by the Church on account of his holding on to Copernican beliefs. And, before Galileo, Bruno was burnt at the stake for reasoned dissent. As recently as a little over hundred years ago, Darwin and Huxley were laughed at by an uneasy Church for saying that man has evolved from 'lower' creatures, and not put on Earth as an act of creation.

The Extent to Which Science is Compatible With Religion

We must, to begin with, recognize two aspects of religion: the ethical aspect and the dogmatic aspect. As regards the ethical aspect of religion, there is an element of universality in it. All religions teach essentially the same basic ethical principles. "Thou shall not kill", or "Thou shalt love thy neighbour", and so on, are not the edicts of just one religion. It is the initially empirical recognition of these values and their slow but sure refinement over the ages consequent to experience and increasing acquisition of knowledge, that has been the common heritage of all humanity. More important, the ethical aspect of all religions is essentially compatible with science. We are beginning to realize today, that there might indeed be a biological basis for ethics: that is, there might be objective assays to determine whether a certain value system, a certain action, a certain behavioural pattern, is "right" or "wrong". The assay will probably emerge out of an increased understanding, in depth, of biological evolution following Darwinian selection, and delineation of behavioural patterns that should have provided an evolutionary advantage. It is, for example, being recognized that altruism may have been an evolutionary imperative. If altruism were not built in our genes, we would probably have been extinct! We are thus beginning to recognize a scientific basis for values which we have so far considered axiomatic on the basis of experience. In fact, a major source of today's social conflicts—for example, the generation gap—has been the belief that all values are eternal. We now know that this is not true. A value which might be desirable today, may not be equally desirable a hundred years from now, when life-styles change. Once we have a scientific assay for determining whether a particular value system is desirable or not, built in that assay will be the prescription for determining its validity at a given time or occasion. Such a "value system" will obviously be much more desirable than the conventional sayings about morality, about what is right or wrong, that have come down to us through the word of mouth, or through rigid convention, custom, tradition or religion.

Incompatibility of Science and Religion

On the other hand, the dogma of a particular religion is non-universal and religion-specific. Since dogma is an inseparable part of a religion and gives it its identity, in common parlance a reference to religion means primarily a reference to dogma of that particular religion. The dogma of all religions is totally incompatible with science.

As already mentioned, the existence of the supernatural—that is, something which is beyond the laws of science—is implicit in religion, no matter what definition one accepts. To add force to it to make it more acceptable, there is also provision in all the religions for the supernatural to take the form of what appears to be natural. Thus, Mesiahs or *Avatars* are born on this earth, and God takes the shape of man or even other creatures, as is supposed to be the case with some of the incarnations of Vishnu. It is this inherent belief underlying religion that has led to the emergence of various forms and shapes of godmen—be it Mahesh Yogi, Satya Sai Baba, Rajneesh, or what have you. These godmen attempt to make others believe that they (the godmen) have supernatural powers which cannot be understood by other men, and that their statements and actions must, therefore, be accepted by others without questioning. Science, on the other hand, does not accept the existence of a high priest, a godman or any other authority that cannot be questioned. In fact, science denies the existence of the supernatural and of miracles which are the very essence of religious dogma. One often witnesses or hears about events which, in the opinion of those who are religious, can have only a supernatural explanation—that is, an explanation outside the scope of the scientific method. In the view of science, all such events—assuming they have ever occurred (which, at times, is doubtful)—do have a scientific explanation, often simple and ingenious.

Religion is based on revelation. Indeed, revelation is the method of religion. Truth was revealed to, and not discovered by, all the religious leaders of the past—be it Moses or Mohammed, Christ or Ramakrishna Paramahansa. The method of science that the scientists use, has no place for revelations of that kind.

Another important attribute of science is that it allows one to make testable predictions on the basis of close observation and collation of available information. Man's landing on moon was a grand experiment involving an enormous number of predictions, where even a single wrong prediction would have spelt disaster. Mendeleev's prediction of the existence of elements and their properties and their subsequent discovery; Pauli's prediction of neutrino; Murray Gellman's prediction of omega minus, a fundamental particle; and the prediction of the existence of planets Neptune and Pluto, are some such predictions that came true and, today, stand testimony to the validity of the knowledge that made these predictions possible; on the other hand, in the entire history of religion there has not been a single such prediction that has subsequently come true.

In science all truths are truths by consensus that is reached among people who are knowledgeable in the area concerned and have formed their opinion by using the method of science and verifying the results personally, or satisfying themselves adequately about the validity of the experiments and of the logic which led to the particular truth. On the other hand, religious truths represent an opinion usually of one religious leader, at most of a few. Moreover, these opinions are rigid. Changing them implies establishing another religion, or at least a sect. Therefore, a given religion, by definition, is static, unlike science which is dynamic and changes with time as more and more evidence comes forth.

It is often said that science progresses by disproving. At least two Nobel Prizes were awarded for discoveries which were subsequently proven to be incorrect. However, in both these cases the persons concerned (H Wieland, S Ochoa and A Kornberg) deserved to receive the Nobel Prize because, had they not made their discovery, the truth as we know it today would not have been discovered at the time it was. Science is, therefore, evolutionary, which religion is not. The growth of scientific knowledge is a continuous process. A religion once founded continues substantially unchanged.

Science has a built-in corrective which takes care of human fallibility on a continuing basis, that religion does not have. In science, a new theory must explain all that was explained by the old theory plus something that could not be explained by the earlier theory. The new theory, in addition, should be capable of making predictions which could not be made on the basis of the old theory, and some of these predictions should, indeed, have been tested and turned out to be right. For example, Einsteinian physics made predictions which Newtonian physics could not, and explained events and phenomena which the earlier physics could not.

Thus, the interconversion of mass and energy, the bending of light in the presence of a large gravitational field, the existence of black holes, and the dependence of the mass of an object on its speed, were all predicted by Einstein and later on substantiated. None of these predictions was possible on the basis of Newtonian physics. That is why we consider Einsteinian physics an improvement over Newtonian physics from which it actually evolved. Contrast this situation with that obtained in religion, where no religion can be said to be an improvement over any earlier religion. If you say something to the contrary—that one religion is an improvement over another—you might initiate a riot!

All new knowledge in science must be consistent with known and established observations. On the other hand, religious dogma (including the so-called miracles, for example, the materialization of objects by the wave of one's hands) is often inconsistent with known and established observations.

Science progresses through modification of a part of the existing knowledge and not by the replacement of the entire body of the existing knowledge. A new religion, on the contrary, often attempts to replace fully the existing religions.

Another important difference between science and religion is that while science is forward-looking, religion is backward-looking. For example, for the followers of science, the more modern the text, the better it is. On the other hand, religious texts on which the followers of religion depend, are generally ancient. In the case of science, the scientists of the present time matter the most; in the case of religion, the founders of the religion who lived in the remote past matter the most. For the followers of science, the events of today and the likely events of tomorrow are the events of the greatest concern; for the followers of religion, the religious events of the past are the events of the greatest concern. The techniques used in science keep on improving with time, and the impetus for this improvement comes from within the framework of the method of science. On the other hand, religious customs and practices do not basically change with time. Whatever changes are brought about are due to forces external to the religion—such as science itself.

An important attribute of science is the right to question. Knowledge advances and science progresses because people exercise their right to question. By contrast, religion demands an unquestioned acceptance of its tenets and dogma. If you question, it must be only to seek clarification and not to doubt.

A scientist can say without any feeling of guilt or shame, 'I do not know'. It would be disastrous for a religious leader to say 'I do not know'; he would simply lose his following. By definition, he knows all! Every major religious leader of the past—the founder of every religion—had answers to every question that one may ever ask. Science would consider such a claim as hypocrisy and deceit. Science takes the unknown as a challenge; religion often leads to a fear of the unknown.

Another important difference between science and religion is that while science (unlike technology) is truly international, religion is not. Scientists all over the world use the same method, that is, the method of science. They employ the same techniques, use the same materials, and publish frequently in the same journals. They are increasingly beginning to use the same language—that is, English—and they form a truly international community in which the professional links are at least as strong as any other link. Contrast this internationalism of science with the parochialism of religion. There are many religions and they differ from one another in many respects. The activities of a particular religion are carried out in isolation of the other religions: in fact, people of other religions are often prohibited from participating. There is little communication between various religions and, therefore, no common language. Religious customs and practices differ enormously, often fundamentally, from religion to religion. Religion, in fact, divides people while science unites them.

Let us look at some specific examples of contradiction between science and religion. Today, we understand reasonably well, what might have been the likely origin of the universe. It is generally accepted by scientists that the universe came into existence about 13 billion years ago, and they can trace the history of the universe backwards to nearly 10^{-42} of a second just after the event of its formation—called the 'big bang' by the astronomers. In this scheme, there is no need to postulate the existence of god, as one must do in religion.

Today, scientists can say with considerable certainty that life on our planet evolved from non-living materials. After the formation of the earth, complex chemical substances were slowly formed from the simple chemical substances that were contained in the primordial atmosphere; such a 'chemical' evolution eventually dovetailed into the biological evolution that led to man. On the other hand, all the other religions demand the acceptance of the belief that man (and, in the case of some religions, other forms of life as well, including women) were put on this earth by God through a deliberate act of creation.

Some of the religious leaders of the past were supposed to have been born through immaculate conception—an idea which is utterly incompatible with the scientific truth about reproduction. And virtually every religion postulates some kind of life after death. The concept of soul is common to all religions. On the other hand, a scientist may ask the question: "Where has the soul been if you can bring a dead man back to life?" as indeed you do when you take out the heart of a person and replace it with the heart of another person.

Secularism and Scientific Temper

Scientific temper arises out of the practice of science, where the method of science is the only tool for acquiring knowledge. Scientific temper leads to skepticism of all truth based on faith and, therefore, leads to a negation of religion that is implicit in the definition of secularism. Therefore, without doubt, scientific temper is an indispensable ingredient of a secular outlook.

Secularism in Our Country

The strength in this respect in our country is the declaration in its constitution that India is a secular state. The State in our country has no religion. Even though religion continues to hold sway over a large number of our people, the redeeming feature is that, in many areas and on many occasions, religion has played no part in public affairs. For example, a large number of appointments that have been made or continue to be made in the country to various positions, are made on a secular basis. There is an increasing tolerance of inter-religious marriages. And we have a substantial number of highly vocal and influential people who are truly secular in their outlook—just as there are a large number of those who are just the opposite.

What about our weaknesses in this regard? One of the greatest of our wearnesses has been that the practice of secularism has been often identified or equated with tolerance. Secularism as practiced in India has been far from negation of religious dogma. It has not even been real tolerance or equal regard for other religions; it has been, in fact, a respectable licence for practicing and propagating one's own creed without any bar or restraint. The architect of secularism as practised in India today was one of the most distinguished citizens of the world, Mahatma Gandhi. Ironically, he did not realize that regarding all religions as equal was a contradiction in terms, as the dogma of one religion often stood in direct contradiction to the dogma of another religion. Identifying oneself with one religion totally, therefore, cannot but make you antagonistic to another religion; any claim to tolerance can only be superficial. A multi-religious society can function well, without invoking religion-based emotional responses that can be destructive, only to the extent to which people are willing to give up their beliefs in their religion's dogma. Similarly, teaching religion as a fact of history is one thing, while teaching religion, with all its dogma, as a desirable way of life is another. The former is a part of liberal education; if such a teaching is coupled with teaching of science and the scientific method, one is left free to make one's own judgement and arrive at one's own decision in regard to beliefs; the chances are that if one is thus left to one's own wits, most people would find it difficult to accept most of the religious dogmas that circumscribe our thinking today. On the other hand, if religion is taught as a package which, along with all its dogma, is considered as the desirable way of life, implicit in such teachings is a denial of science and of all other religions, and the talk of equality of all religions in such teachings becomes a mere farce. From this point of view, unfortunately, we have failed.

The irony is that even though we are committed in our constitution to secularism, the greatest failure in this regard has been of the government and the political machinery. For example, the entire concept of reservations as practised in the country today is unscientific and nonsecular. In fact, continuance of such reservations in service after more than four decades of Independence is an indication of the failure of the government and the political machinery in respect of universalising the concept of secularism within the State. In fact, it looks like a virtual conspiracy, for whatever reasons there might be, to keep the Scheduled Castes and Scheduled Tribes very nearly where they have been before

Independence, instead of closing the gap between them and the rest of the people of the country. If we were truly secular, our efforts would have been to provide the under-privileged with at least equal opportunities if not greater facilities, and not to perpetuate inequality in regard to standards through the system of reservations. Today, a person belonging to Scheduled Caste or Scheduled Tribe knows that he does not have to achieve high standards that the others would need to for obtaining a certain position or privilege. Therefore, what would be the source of his/her motivation for achieving excellence? The fact is that, in the long run, real success in life is related to merit and meritorious performance. The perpetuated system of reservation, as we have today, is an insurance against effort which would lead to the development of such merit and, eventually, to meritorious performance. By lowering the standards for Scheduled Castes and Scheduled Tribes for decades after Independence, we are only relabelling them, rather than removing the label. Even the highly restrictive meaning attached to secularism generally in our country would not justify such discriminatory policies.

And then, much of our politics is based on considerations of religion, caste or sub-caste. We still permit religion-based organizations and political parties, and the right to religious teaching in minority institutions; our textbooks often have a religious bias. Religious leaders have, often, an important say in public matters, and we thus allow Shankaracharya of Puri to get away unscathed in spite of his open and vocal support to Sati. We still willingly consign crores of rupees to fire to appease our gods, be it for rain or political benefit. Most of our holidays are religious holidays. Even state functions have religious overtones; for example, government buildings are lit on Diwali day. We provide indirect support to religion by supporting institutions that are founded and run by religious fanatics or by those (such as the godmen of our country) who practise large-scale public deception. And in spite of more than four decades of independence, and commitment to secularism, we still do not have just one law for all the people of our country. We let our laws be significantly determined by religious considerations. Muslims can have more than one wife, and Sikhs can carry *kirpan* on the basis of their personal law. We still often require people to identify their religion in our application forms, and our political leaders openly and publicly participate in and encourage religious practices, rather than secular ones. The situation is made worse by the support and publicity of such practices by our media. There has been much unauthorized encroachment of public land for shrines and religious practices, that has been condoned by the State.

Therefore, it is no surprise that the first identity that people seek in our country even today, is an identity based on circumstances of birth, such as religion, language, the State they come from, or the caste or the sub-caste they belong to, or their social status. The national identity or the identity as a citizen of the world gets relegated far into the background. We have not realized that poverty and communal or religious identity go hand-in-hand in the world of today. Religion is the opium of the poor. What has not been understood is that secularism is their salvation.

Why this situation? If one were to record one important reason, it would be the lack of scientific temper. And scientific temper is not something that you inculcate overnight. It is something that grows on you and grows with you. It is something that we imbibe from the environment—the environment in the house, in the family, and at school. Unfortunately, such a climate does not exist in our country. Indeed, this has been one of our greatest failures. India cannot be secular without its people having a spirit of secularism. This can only be achieved and sustained in an environment of scientific temper.

Today, divisive forces in our country have become the greatest single threat to our integrity, unity and, in fact, our very existence as a sovereign nation. The most effective weapon we have to fight these forces with is scientific temper. The sooner we learn to do so, the more assured would our future be.

Under these circumstances, is it enough to merely state that the State has no religion and, thereby, lay a claim to secularism?

XX

MODERNITY AND SCIENTIFIC TEMPER

P M Bhargava

I feel deeply honoured to have been invited to deliver this lecture in commemoration of the birth centenary of Jawaharlal Nehru. I was privileged to have met him on several occasions, each of which is deeply etched in my memory.

The subject on which I have been asked to speak, Modernity and Scientific Temper, is particularly important as we are committed to both as a nation. Perhaps, to Jawaharlal Nehru, a discussion on modernity, or on scientific temper would have seemed redundant, for to him there was no questioning either of these. He said, "The applications of science are inevitable and unavoidable for all countries and people today. But sometimes more than its application is necessary; it is the scientific approach, the adventurous and yet critical temper of science, the search for the truth and new knowledge, the refusal to accept anything without testing and trial, the capacity to change previous conclusions in the fact of new evidence, the reliance on observed fact and not on preconceived theory, the hard discipline of the mind—all this is necessary, not merely for the application

---

* This article is based on the Nehru Centenary Lecture given at the University of Delhi in August 1989.

of science but for life itself and the solution of its many problems". But, the unfortunate fact is that today, on the political scene, we first use words to impress and woo support, and then give these words a meaning which has no relationship to the real meaning of the word. Thus, there is no political party in the country that would be prepared to say that it is not wedded to socialism, secularism and democracy. There would be no individual who will accept that he does not have scientific temper. Even the most rabid communalist or communal parties call themselves secular, and the most committed revivalists swear by *modern* India. It is in this context that a discussion on modernity and scientific temper in the same breath becomes relevant, particularly at this occasion which I regard as an occasion of stock-taking of the state of the nation.

There are two ways in which the practice of science and the knowledge generated through it, has increasingly influenced man and society: (1) through providing material gain, comfort or convenience of one kind or another, or material means of acquiring power, or improving the quality of life; and (2) through influencing our thought processes, views and value system, and the manner in which we arrive at decisions or form opinions—including decisions or opinions that are related to practical application or use of scientific knowledge. Scientific temper is the quality of mind that defines and determines the latter aspect of relationship between science and society.

The Origin and Meaning of the Term, "Scientific Temper"

Jawaharlal Nehru was the first to use the term, "scientific temper"; it was a phrase much in Nehru's vernacular. He reiterated it not only in speaking of science, but also in exhorting his countrymen in diverse contexts. The phrase is an attractive one and has both brevity and comprehensiveness, for temper indicates all the hues of man's thinking, nicely qualified to the plausible and rational with the adjective scientific. It implies a willingness to consider all facts, and not merely facts "which are in consonance with one's own thinking or comfort it." Going further, it means an active search for such information by study and questioning. It also implies a trust that events are shaped by the fruits of man's labour, and a healthy scepticism towards all claims of supernatural participation in his affairs. In fact, the scientific attitude (or temper) is simply one of an adherence to facts, an ability to revise opinions, and a rational scepticism to claims for non-material intervention in our affairs. The spirit of inquiry and the acceptance of the right to question and be questioned are fundamental to scientific temper.

In 1981, about 30 amongst India's well-known individuals, representing a variety of professions, signed a much discussed and publicised Statement on Scientific Temper which was published subsequently by the Nehru Centre, Bombay. To quote a part of this statement:

"Scientific temper involves the acceptance, amongst others, of the following premises:

(a) that the method of science provides a viable method of acquiring knowledge;
(b) that human problems can be understood and solved in terms of knowledge gained through the application of the method of science;
(c) that the fullest use of the method of science in everyday life and in every aspect of human endeavour from ethics to politics and economics, is essential for ensuring human survival and progress; and
(d) that one should accept knowledge gained through the application of the method of science as the closest approximation to truth at that time, and question what is incompatible with such knowledge; and that one

should from time to time re-examine the basic foundations of contemporary knowledge."

It could, therefore, be surmised that scientific temper implies the acceptance of the premise that the method of science is the only method through which knowledge may be acquired, and of the premise that all human problems can be (and should be) solved only in terms of knowledge acquired through the application of the method of science. The scientific temper, therefore, requires rejection of all that is incompatible—at any given time in history—with knowledge acquired through the application of the method of science.

To fully comprehend the implications of scientific temper in relation to our obligations and duties, we must, therefore, first understand what the scientific method is.

What is the Scientific Method?

We are all aware of the benefits of science to humanity, but we seldom recognise that they are the result of a simple, systematic, well-defined, and objective approach: the application of the method of science to solution of problems and discovery of truth. The seeds of the scientific method were sown by Roger Bacon in the thirteenth century; it was further embellished and stated in its present form two centuries later by Francis Bacon. It has four distinct steps—the framing of the question, the framing of a hypothesis, the doing of an experiment, and arriving at the answer which may be a fact or a generalisation in the form of a theory or law. The question arises out of careful discussion or careful analysis of existing information. The hypothesis is a possible answer; a hypothesis is valid in the scientific method only if it is testable. The experiments must be done with care, without bias and prejudice, and every observation recorded irrespective of what one wished it to be;

unexpected observations have often led to major scientific discoveries. The method of science rejects revelation as a means of discovering truth and substitutes it by the technique of observation, followed by careful experimentation and logical deduction; it, therefore, stands in direct contradiction to the ways of religious dogma and faith which are based on the premise that truth can only be revealed.

The following are some of the important attributes of knowledge gained through the method of science:

- It is compatible with observation and insight, reasoning and intuition, systematic work and creative impulse.
- It is not dogmatic or unreasonably insistent.
- It does not ask you to accept anything on faith.
- It does not regard truth as unchangeable or unmodifiable. It considers knowledge as open-ended and ever-evolving.
- It lays emphasis on verifiability and repeatability wherever possible, and on the fact that scientific theories, laws and facts allow one to make predictions which can be tested.
- It has no "high priests" who cannot be questioned.
- It possesses no articles of faith, has no prejudices, is not orthodox and is not conservative.
- It is universal.
- It is forward-looking.

It would be obvious that in all these respects scientific knowledge or truth, is an antonym of "knowledge or truth" that derives from religious or other dogma. (Religions—all religions—have two aspects; the moral and ethical, and the dogmatic. Moral and ethical dictates of all religions have a very large overlap and are, by and large, consistent with the value system generated by the practice of science. This is, however, not true of dogma which is different from one religion to another and gives a religion its identity. Therefore, identification of an individual with a particular religion implies identification with the dogma of the religion.)

Acceptance of the above-mentioned attributes of knowledge is, as has already been mentioned, an important manifestation of scientific temper.

Modernity

Modernity has a wider connotation than modernization. Modernisation refers to civilization and mainly implies a high level of literacy and urbanization with better standards of living and a high per capita income. Modernity, on the other hand, connotes a certain type of culture whose quality is determined by rationality, a liberal spirit, plurality of opinion and of centres of decision-making, autonomy of the various fields of experience, secular ethics, and respect for the private world of the individual. Although modernity leads to industrialisation and modernisation, modernisation itself does not necessarily imply the growth of modernity. Let us look at some of the concommitants of modernity (and modernisation).

Incursion of Science and Technology into Society

Modernisation, per se, is the incursion of science and technology in our daily lives. It is not only in regard to matters of luxury that the impact of science and technology has been felt in our everyday life in this century. It is equally so in regard to what are universally recognised as basic necessities: food, clothing, housing, health, education, employment and social justice.

Today's food requirements around the world would have never been met even to the extent to which they are, without using the knowledge we have acquired during this century in regard to genetics. New techniques in genetic engineering provide the major hope for taking care of the food requirements of the increasing population of the world in the coming years. The use of synthetic material for producing items of clothing in the last four decades has increased to an extent that in many parts of the world it would be difficult to find clothing having no synthetic yarn at all. Similarly, the use of non-conventional material for constructing houses and other buildings has increased dramatically during the last few decades; many of these materials were totally unknown before the last war. Today, cure for a vast variety of diseases which were considered incurable only forty years ago, are known; we have eradicated from our planet diseases such as smallpox which used to be a major scourge of humanity. Television and computers have brought revolution in education in many countries of the world. Science-based industry has opened up totally new avenues of employment.

The shift from agricultural economy to industrial economy that we have seen in many countries of the West and which is inevitable, sooner or later, in other countries as well, has been largely brought about as the result of a revolution which, on one hand, allowed mechanisation of agriculture and increase in agricultural productivity through use of agents such as fertilizers and pesticides and, on the other, provided employment to those who no longer needed to be employed in agriculture.

The need for framing new laws that will govern the use of space, or artificial insemination donors, surrogate mothers or new life forms—and a host of other new discoveries which were not dreamt of just two decades ago—is being increasingly felt around the world. Today, in some cases the only way to establish or disprove paternity without doubt may be the technique of DNA fingerprinting which was unknown a decade ago.

By transforming the modes of conducting war through provision of totally new kinds of armament and new techniques of warfare, science and technology have transformed the art and science of politics. Someone who is not knowledgeable about the scientific basis of the nature and structure of the arms industry around the world is unlikely to make a good political leader irrespective of the colour or hue of his party. And by bringing out the contradiction between the irrational elements of religion and the rationality of science, science has acted as an instrument of a major change in our perception of religion and its role in everyday life.

For the first time in human history, the outcome of scientific research and technological development has had a significant impact on other forms of creative activity such as painting, music, films, even literature; modern science and technology have provided not only new materials but also new concepts, and engendered new attitudes, in these areas.

Indeed, the changes that have occurred in the scenario around us and in our life styles during this century far outstrip, both qualitatively and quantitatively, the changes that occurred in this respect in the entire period of human history till the end of the last century. This change in the face of our planet and in our life styles during this century has been, in a very large measure, due to the change in the face of science and technology during this period. It is this intermeshing of science into every aspect of our life that has been one of the main assets of modernization.

Secularism

Another dramatic change that has occurred during this century is an increasing loss of the hold of religion over the State—even over the minds of people. There is no question that people around the world have become more secular in their outlook. (By secularism I mean here the definition given in Longman's Dictionary of Contemporary English, "a system of social teaching or organisation which allows no power to religion or the Church", or in the Oxford Dictionary where "secular" is defined as "sceptical of religious truth or opposed to religious education".) Towards the end of the first half of this century, my grand-uncle was the professor of physics at one of our prestigious universities of that time. However, in our house, my grandmother would not eat or allow any one in the house to eat in utensils that were used to serve anything to a non-Hindu. Such a situation which was extremely common at that time, is unlikely to be obtained today in the house of any University teacher anywhere in the country. When I was thirty years old, and I say this from personal experience, inter-caste marriages were extremely few—leave aside interreligious marriages. Today the number has increased by orders of magnitude, to an extent that no one takes any particular note of such a marriage today. Secularism, therefore, is an important concomitant of modernisation. If you are not secular, you simply are not modern in your outlook.

Socialism

Perhaps, the most important aspect of socialism is the denial of the exploitation of one individual or group of individuals by another individual or group for personal or group gain. Since money is the most common instrument of such exploitation, socialism demands a system where money in the hands of one individual cannot be used to make more money for that individual, implying that all the money that one may obtain in a socialist society would be related directly to the effort or labour put in by the individual. Although, as a concept, socialism has been with us for well over a century, its practice began with the October 1917 revolution in the Soviet Union. Today, it is a highly respectable word—so much so that it will be difficult to find anyone who would be prepared to say that he is against socialism; the only difference between individuals would be in respect to the meaning that they attribute to the term, "socialism". Thus no civilized man on the face of our planet today will say he is in favour of exploitation of one individual by another; the only difference between people would be in respect of what they consider as exploitation. Perhaps, the lowest common denominator of all definitions of socialism is one kind or another of social control—that is, control by the people, irrespective of their status in the society from the point of view of money, riches, position or power. This concept of social control is a modern concept.

Democracy

Democracy defined as "Government of the people, by the people and for the people", as Abraham Lincoln said—that is, Government by elected representatives of the people—has been, no doubt, practised for well over two centuries in parts of the Western world. But at that time democracy at one place was sustained by colonial rule and exploitation at the other. It was, therefore, in truth, pseudo-democracy. Democracy came into its own only with the disappearance of colonial rule. Today, there are only a few countries where there is no Government of the people, by the people and for the people, that is, Government of appropriately elected representatives, at least in name. A country which does not practice democracy—at least the above-mentioned lowest common denominator of democracy—cannot today call itself civilized. (From this point of view, the white South Africa is, without doubt, the most uncivilized country in the whole world today!) The process of democratisation has been a hallmark of modern times.

Human Rights

It does seem incredible but the fact is that the concept of basic and universal human rights is a relatively new concept.

It was only after the last World War that United Nations adopted the Charter of Human Rights which include the right to: life, liberty, security of person; equality before law; travel; freedom of thought, conscience and religion; formation and expression of opinion; equal access to public service; social security; work; rest and leisure; health and education.

Today, only that country that guarantees basic human rights, or at least makes the utmost possible effort to see that these rights are secured for its citizens, alone has the right to call itself modem in outlook. Slavery and bonded labour are totally incompatible with modernity.

Peace

The concept of lasting world peace is essentially a modern one, It is only in the present half century that, for the first time, the rank and file around the world have begun to recognize the need of lasting peace, and that peace is the only insurance for continued development. The optimists amidst us—and I count myself as one of them—feel that there is no way out of lasting peace, that sooner or later people of the world and their governments would realize that war is not the way to solve conflicts and that even major conflicts can be resolved through peaceful means.

The One-world Concept

For the first time in the history of man, we have begun to realize that the future of all mankind is closely interlinked. Failures of any kind—be they in respect of provision of basic needs, or in respect of fruits of modern technology that have brought vital changes in the life styles of those who can afford them—anywhere, in any part of the world, are bound to have repercussions everywhere else. Islands of affluence are no longer tenable—certainly not for long. In this statement is built a statement of responsibility of the more affluent nations towards those who are not so affluent, and the need for *sharing* for the sake of one's own security and continued prosperity. This concept is less than four decades old.

We have thus realized that nuclear war between any two nations is bound to spell disaster for all of humanity; the same would be true of environmental degradation. Use of chemicals such as those continued in certain aerosols, that would deplete the ozone layer that protects our earth from harmful radiation from space, is bound to have global repercussions. And there cannot be one, just a *national* policy for disposal of certain kinds of waste such as nuclear waste; for any policy to be viable in the long-range interest of the country itself, it must take into account the international implications.

If the developed countries wish to ensure good health for their citizens, it is imperative that infectious diseases are wiped out from the whole world including, of course, the developing countries. If, for example, there were an outbreak of cholera in the United States, it may take far more lives than an outbreak of cholera at the same scale in India where people have natural immunity due to being exposed to small dosages of the toxin in their earlier life. We have seen something like this happening only recently in regard to AIDS. Countries which had no case of AIDS and which have been far away from the Western world where AIDS was first noticed and spread, are reporting an ever increasing number of cases all the time.

There is no country in the world today which is totally self-sufficient. International trade is a hallmark of modernity. An awareness of global accomplishments has become essential for setting one's own objectives on the national scale. An awareness that the destinies of all men are interlinked and that every human being has certain inalienable rights which we must respect and which are not negotiable under any circumstance, has become an integral part of contemporary, modern civilisation. Those who do not possess this awareness can, today, call themselves at most second-class citizens of the world.

Information

For the first time in our history, we have begun to recognize the right to information as a basic right of all humanity. We are beginning to realize that information is necessary for acquisition of knowledge, just as knowledge is necessary for wisdom. Knowledge hierarchy has now begun to outstrip all other, conventional hierarchies that were, for example, based on considerations of power, money, status and various circumstances of birth, such as religion, caste, sub-caste, nationality or language.

Communication and travel

One of the greatest revolutions that has ever occurred in human history has been the revolution in regard to communication and travel which has taken place in the last few decades. In fact, communication and travel have become one of the most important sources of acquiring information, knowledge and wisdom. The ever-increasing democratisation of the facilities for communication and travel has become virtually synonymous with modernity.

Anti-obscurantism

A fear that has sustained obscurantist beliefs—that is, beliefs that are conceived in irrationality, unreason and ignorance—is the fear of the unknown. As the repertoire of human knowledge has increased—and it has increased exponentially over the last few decades—many areas of erstwhile ignorance which provided a fertile soil for the sprouting of obscurantist beliefs, no longer exist: the unanswered questions have been answered. Since truth is only one whereas untruths can be many, obscurantist beliefs had divided people. With their disappearance and with the emergence of the truth, the foundation of such beliefs has been shaken. For example, the strongest argument against apartheid is the genetic argument—that in any outbred population, all abilities get randomly distributed so that no group would be justified in considering itself superior to another group on the basis of merely the circumstances of birth or the colour of skin, provided, of course, the group size is large as has been the case in South Africa.

Relationship between Scientific Temper and Modernity

While on the one hand, the practice of science generates values (such as altruism and honesty) which we have, through history, axiomatically and universally accepted and which have stood the test of time, on the other hand, the practice of science and the scientific temper have also provided the basis for the development of the *modern* values we have talked about. As we have already seen, science by definition is anti-obscurantist and there is one and only one science. Science is, therefore, a major unifying force in the world today, the scientists forming a truly international community in which the professional links are at least as strong as any other link. The pace of development of science today is so fast that there is a constant need for increased communication and travel. In fact, the scientific community today could well be one of the most travelled groups barring, of course, the airline pilots and other staff. We have also seen that science is democratic though not in a trivial way. And the right to question is a fundamental right in science.

Science is truly secular: there is no separate Chinese science or Ethiopian science. Science argues totally against distinctions made on the basis of circumstances of birth. This view is sustained by the fact that science has not flourished in societies where any such discrimination has been or is being practised. The status of southern states in the United States even today, in respect of science, is far lower than that of the northern states. (Civil war in the United States was fought between the northern and the southern states on the question of slavery which was practised in the southern states but not in the northern.) There is very little worthwhile science that has come out of South Africa in comparison to the countries of the West that have almost the same population as that of white South Africa; most, if not virtually all distinguished scientists that were born and brought up in South Africa have left South Africa. With the solitary exception of Israel (for very special reasons), there is no country in the world dominated by religion where science has flourished in modern times. In fact, the pre-eminent position of India amongst the developing countries in respect of science and technology is no doubt substantially due to the secular nature of the country; it has the largest number of Hindus in the world but it is also the second largest Muslim country in the world, and has more Christians than all of Australia. And the strongest argument today in favour of peace as the only insurance against the total destruction of mankind, is the scientific argument.

Science does not accept class distinctions based on circumstances of birth and is non-exploitative; it, therefore, lends total support to socialism. Further, beginning with Karl Marx in the last century, science, has contributed significantly to the formulation, development and practice of political ideologies and economic theories. Most of the outstanding scientists of today tend to be "left of the centre" as they find a familiar echo in the ideologies of the left. A 'right' reactionary scientist would today be regarded as a contradiction in terms.

By bringing out the contradiction between the irrational elements of religion and the rationality of science, science has acted as an instrument of a major change in our perception of religion and its role in everyday life.

By providing a rational and objective basis for decision-making and for the equality of all human beings irrespective of caste, creed, sex, age or nationality, by making travel and communication as fast as it is today, by removing drudgery from everyday life, by providing new forms of entertainment, and in many other ways that are in consonance with modernity, science, has acted as one of the most important determinants of social change. And the knowledge acquired through the advances in physics and biology has contributed to the creation of a climate in which equality of all men comes to be better recognized, and a ground is prepared for equitable social justice.

Science is also beginning to provide a rationale for determining whether or not a particular value is in the long-term interest of mankind or not. Concepts such as those of basic human rights, derive their maximum support from what emerges out of knowledge gained in areas such as biology and physics in the last few decades. If a brahmin can accept blood transfusion from a non-brahmin or an untouchable, what would be the basis for his opposition to the marriage of his son to the daughter of the untouchable on grounds of caste (or lack of it)? In fact, modern biology has provided a framework for a major revolution in human thought in this respect; for example, as has already been mentioned, the strongest argument for secularism or against racial discrimination, is the biological argument.

There is also little doubt that the various roles of science will get increasingly consolidated, and the process of making decision about science would become even more closely intermeshed with the process of taking socio-politico-economic decisions than it is today.

Science has thus been an important and indispensable tool of modernisation in every form or respect. And science can only flourish in a society in which scientific temper prevails. *Scientific temper is to modernity, what science is to modernisation.* The values that are inherent in scientific temper are the foundations of modernity, just as science is at the base of modernisation. Claims to modernity without scientific temper must be considered sterile. It is in this context that we must now look at the state of affairs in our country in this regard.

Situation in India

Development of scientific attitude among the people was an important part of Nehru's vision of India. He recognized, however, the magnitude of the transformation required of contemporary Indian society before his vision could materialize, and this was sufficient to despair even an optimist like him, for scientific temper is conspicuously lacking in the country, even among those with an ostensibly scientific training.

India has had many rich traditions but not of objective, rational and scientific thinking. India never went through the renaissance that brought up and nurtured the rationalist tradition in Europe. The industrial revolution that had established itself in Europe for well over a century, had not even touched our country when the independence movement started. On the contrary, the latter became strongly bonded with revivalism, in no small measure due to the personal beliefs of one of the most remarkable man of our times, Mahatma Gandhi. There can be little doubt that by this means he was able to mobilize enormous numbers to the national purpose, but he also set in motion various factionalisms, which remained submerged during the intense drive for independence but have raised their ugly heads since then. Nehru represented a positive gain for independent thinking, setting himself up against all types of reaction. One remembers with gratitude that he was the only national leader who had ridiculed the absurd Ashtagraha episode. But the impact of his ideas in this regard on the country as a whole was probably small. While it may have set at bay some obscurantist and religious forces, these were only biding their time; their resurgence was markedly noticeable even in the brief regime of Lal Bahadur Shastri. Ever since, obscurantism has been on the rise and, quite shamefully, has received support from quarters whose primary responsibility should have been to check obscurantist ideas: unfortunately, the educated, the academicians and the scientists of today are as obscurantist and superstitious as anyone else.

Just look at the followers of the godmen of our country such as Satya Sai Baba, Mahesh Yogi or Rajneesh, or visit the house of anyone belonging to the privileged categories, and you would be shocked to note the irrationality and unreason that pervades their life due to their obscurantist and superstitious beliefs. It may be difficult to find a minister who does not have an astrologer, and it is not unusual for a godman to have a say in political decisions. Scientists lay aside the mantle of incredulity and deductive logic when they get home and kick their shoes off, relaxing into every kind of obscurantist fad and fallacy. Doctors still see no contradiction in their patients visiting Tirupati or the local temple for cure of physical ailments; indeed they do so themselves. Several scientists have little conviction in regard to the benefits of inorganic fertilizer in their kitchen gardens; fertilizer is something to be doled out to the farmer, or used in speeches. Daily newspapers and magazines yield rich dividends to the seeker after obscurantism; practically every newspaper has an astrological corner with predictions for the week to follow. The child of a dentist will still get married in the small hours of the morning so as not to offend the planets in their whizzing courses. There are auspicious days for travel, and specially auspicious, individualized gems and numbers. In a poor country where millions live below the poverty line, a vast amount of wealth is consigned to *havanas and yagnas* (Hindu religious practices resorted to in adversity to propitiate gods including, for example, the rain god!)

Therefore, as was said in the earlier-mentioned Statement on Scientific Temper,

> Despite Jawaharlal Nehru's advocacy of scientific temper, we are witnessing a phenomenal growth of superstitious beliefs and obscurantist practices. The influence of a variety of Godmen and miracle makers is increasing alarmingly. The modern tools of propaganda and communication are being used to give an impression that there exist instant and magical solutions for the problems that confront our people. In an age when man has travelled to the moon and returned safely, astrological predictions based on the movements of planets, or on the lines of one's palm or the number of alphabets in one's name, are widely believed. Food fads and irrational health practices are on the increase. Myths are created about our past. The origin and rôle of the caste is explained in a way that would justify it and imply that some castes are inherently superior. The ancient period of our history is interpreted to inculcate chauvinism which is false pride; the medieval period is misinterpreted in a way that would fan communalism; and the struggle of our people for freedom is over-simplified as if it was the handiwork of a few great leaders and the masses of our people did not matter.

Decisions are taken, and opinions formed, today in our country, on the basis of bias or prejudice, or inadequate information. Little respect is shown for facts. Rarely, if ever, are all the possibilities taken into account or the 'enumerations complete' in accordance with Descartes' Four Rules of Logic in his *Discourse on The Method*. Future of many is staked for a short-term gain of few. Historical imperatives are rarely recognized—and if recognized, are ignored. And all this is as much true of the individual as of the community, as much true at the social level as at the political. Admittedly, these are worldwide problems but, in comparison to the more developed countries, they are an order of magnitude greater in India where they particularly stand out against the country's many remarkable successes since Independence. A vast variety of even our scientists are no exception to the common 'rules' of our society today that argue and militate against the scientific temper.

Our Obligations in the Birth Centenary Year of Pandit Jawaharlal Nehru

Promotion of scientific temper is now, according to our constitution, one of our duties as citizens. The only permanent long-term insurance for getting rid of obscurantism from our society is to lay emphasis on the right kind of education. One of the professed objectives of education must be to equip the recipient to fight obscurantism of which he would otherwise be a victim. Education, right from the beginning, should be science and knowledge-based. The time has come when we must realize that if any action based on the dictates of religion, classical philosophy, dogma, custom, convention and tradition, goes against the basic, long-term, legitimate interests of the society at large, those in power must take steps to ban such action. The press and Governmental publicity agencies like radio, television and the State departments of information, *can* be highly effective in fighting obscurantism in the country, but they are not.

Efforts to improve science teaching should be supplemented by efforts to popularise science.

As many voluntary organisations as possible should be formed and supported by people who are fortunate to be emancipated from the shackles of obscurantism themselves.

Popular science movements such as Kerala Shastra Sahitya Parishad (KSSP) must be fully supported. No obscurantist or superstitious activity must be permitted under the auspices of the Government or by those who run it, under the pretext of individual freedom.

Professional persons, public leaders, scientists and other academicians must not only exhibit scientific temper in their professional life but also in their personal life; the dangers of dichotomy in this regard, in terms of its effect on the public, must be recognized.

Conclusions

If one were to pick out three or four most important reasons for the country's backwardness or failure in many areas, the lack of scientific temper would be one of them. Unfortunately, it is a vicious circle. Lack of social justice or of the availability of information on a mass scale, widespread illiteracy and lack of appropriate education (often even amongst those who have a University degree!), vastly restricted communication, and economic inequalities that have led to highly affluent islands in a sea of poverty and want, have been major impediments in the spread of scientific temper; on the other hand, unless there is personal and collective commitment to scientific temper, it seems most unlikely that the above-mentioned problems will be solved.

In this year of celebration of the birth Centenary of Jawaharlal Nehru, it is our obligation to recognize the link between modernity and scientific temper, and to realize that for India not to become an antique in the community of nations but to stay on the forefront of modernity, it is important that the above-mentioned vicious circle is broken. It is *our* obligation to do so, for we have had the privilege of higher education; we have thus a very special responsibility in this regard: it is a debt we owe to the country. And the best way to pay this debt would be through personal example. Isn't that what Jawaharlal Nehru did for us, for the whole world?

Let us hope that the Indian society's resilience and capacity to absorb new ideas, will prove to be an asset in this regard.

XXI

THE SEVEN DEADLY SINS OF THE CLERGY

P M Bhargava

It is a cliché that today man has all the means to destroy all of his species through weapons of mass destruction possessed by many countries led by the USA. There is no country in the world today that can be trusted not to use them; in fact, the larger the stock of the WMDs, the greater is the illusory arrogance of power and, therefore, the greater the chances of the country using them.

The only insurance against something like this happening would be an effort towards one world and one Government that would safeguard the interests of all its constituents: a real coming together of peoples of the world, with cooperation and not confrontation being the buzz-word. Why is it then that this is not happening? One of the main reasons is the baggage we are carrying today of seven deadly sins of the clergy that control religion, be they Hindu, Muslim, Christian or any other. Indeed, the clergy are the second multinationals around the world whose primary concern is to safeguard their interests as against the interests of the people. They are the main—in fact, the primary—stumbling block in the spread of scientific temper which helps unite people.

---

* This article was published in *The Tribune,* 2nd January 2006.

Now the seven deadly sins of the clergy of today:

(1) They survive on misinterpretation of the teachings of the founders of their religion or its other respected leaders. An example would be the emergence of Wahhabi-Salafism in the Islamic world. Today's Islamic fundamentalism and terrorism is a direct consequence of the ideology of Wahhabi-Salafism. The way in which the Wahhabis and Salafis have misinterpreted Quran is extremely well-documented in the book, *Terrors' Source: The Ideology of Wahhabi-Salafism and its Consequences*, authored by Vincenzo Oliveti (a pseudonym of Prince Ghazi of Jordan, one of the most illustrious fortythird-generation living descendents of Prophet Mohammed), published by Amadeusbooks, PO Box 10743, Birmingham, B12 8ZX, UK. The way the Christian clergy have done the same thing with Christianity is clear from today's bestseller, *The DaVinci Code* by Dan Brown. The only people who have gained by misinterpretation of the teachings of the founders of the great religions, are the clergy.

(2) It is the clergy who have invented miracles and attributed them to the founders of the great religions. Miracles have been the single greatest weapon in the armour of the clergy. Would Sai Baba of Puttaparthi ever have the following he has if he was not perceived to be capable of performing miracles? It is another matter that everything that he can do which people perceive as miracles, can be done by scores of others who are (unlike Sai Baba) honest and call it sleight of hand. An outstanding example of deceit implicit in the phenomena of miracles is that of Mother Teresa. I had the privilege of meeting her and was extremely impressed by her humility and humanity which alone should have been enough in any civilized world to confer on her the status of a Saint. However, for her to be canonized, Vatican required that it be shown

that she had performed at least two miracles; therefore, these miracles had to be *invented*, with (almost) everyone knowing that they were actually never performed! It is to her credit that she never ever in her life claimed that she had performed any miracle.

(3) The clergy invented the concept of "divinity" which implies that one's life is totally controlled according to what has been ordained by the "divine" power (whatever that may be), and since the clergy represent this divine power, they and they alone can help you change the course of your so-called "destiny". Most Gods—certainly in the so-called Hinduism—are bribable, and the clergy tell you how and what to give as a bribe. Their professed link with the divine power makes them a closed community. Can you think of a Shankaracharya being a Shudra? The Hindu clergy tell you that your caste is a divine dispensation and defines your duties and obligations—in spite of the fact that no religious leader (past or present) could tell the caste or religion of a new-born child! There are no definitive, fool-proof genetic markers for caste. I have no way of finding out in which caste or religion was my lovely adopted grand-niece, Riya, born. If you are designated entirely on the basis of your parentage, as belonging to the lowest caste or being casteless, you must accept that you have no more right than an (unwanted) animal and that you must do without complaining, all the dirty work of the higher castes, with the only compensation being abuse and insult which the higher caste believe they have a right to dispense.

(4) One of the paramount duties of the clergy everywhere has been to distort history, and to first invent and then present legend as history. For example, common sense tells you that Rama and Krishna, and the stories associated with them, are legends—in fact, fairy tales like those of Hans Christian Anderson or Grimm. Dan

Brown's *The Da Vinci Code* brings out, with courage and elegance, the attempt to distort history in the Christian world. There is no codified religion the clergy of which have not attempted to "sanitize" history to suit their interests.

(5) Science has been the biggest enemy of the clergy—perhaps all through history but certainly from the beginning of Renaissance in Europe from which time organized science began to evolve. Thus Bruno was burnt at stake and Galileo incarcerated for stating a truth arriving at by using the method of science. Opposition to abortion and renewed efforts in the USA to give equal status in school teaching to creation and evolution to explain the origin of man, are other contemporary examples. A major attempt of the clergy all over the world, in every religion, has been to replace evidence and truth by belief and myth. Their preachings have, therefore, been the greatest single impediment to the development of a knowledge-based society in the world, which alone can lead the way to universal peace.

(6) The clergy and their followers have been, in fact, the single largest promoters of war and other conflicts around the world in the last many centuries. Examples would be the Wars of Crusades, the religious conflicts in Ireland and Central Europe, the Israel-Palestinian problem, and the problem between India and Pakistan. The clergy have misled people all the time by giving them a feeling of greatness by simply belonging to their religion or sect. They then subtly convert this feeling into the right to govern others who are not so "great". There are lessons in this process for our management and ad gurus!

(7) The clergy have, all through history, kept their followers bound to laws that often have no basis in reason, humanism or basic human rights. Not only that, they

> interpret the so-called "religious laws" to suit their convenience. Indeed, one of best things that happened to the Hindu community in India in the above context was the codification of Hindu Law. Unfortunately, this has not happened with the one billion–strong Islamic community around the world. The Islamic clergy have tied this community down by various—sometimes conflicting—provisions of Islamic personal law, the Sharia.

My personal commitment is to reason, to basic human rights, to scientific temper, and to evidence-based truth, and not to any religious dogma. Nevertheless, in a democratic world, everyone must have a right to believe whatever one wishes to as an individual, and we must grant that right to every individual. However, no one should have the right to preach to others, using falsehood and deceit, their own beliefs including belief in religious dogma. One of the biggest challenges we, therefore, have today is to eventually decimate the hold of clergy on the people so that they may think freely and on their own. Unless that happens, we cannot dream of a conflict–free world. The time has come when we must start thinking seriously about how to achieve the above objective: that is, to punish the clergy of all religions for their seven deadly sins.

XXII

CONCLUSION

India is the only country where commitment to scientific temper is, by the country's constitution, a duty of its citizens. All of us are, therefore, *obliged* to do our best to abide by it. If and when that happens, the chances of conflict at all levels would decrease.

Questioning when there are sufficient reasons to question (which reasons the method of science states) is a crucial element of scientific temper.

Let us remind ourselves of what Lord Buddha said about questioning:

Believe nothing
Merely because you have been told it
Or because it is traditional
Or because you yourself imagined it
Do not believe what your teacher tells you
Merely out of respect for the teacher
But whatever, after due examination and analysis
You find to be conducive to the good
The benefit
The welfare of all beings
That doctrine believe and cling to
And take it as your guide.

Printed at : Magic International Pvt. Ltd., Greater Noida